The American Ceramic Society

1 0 0   Y E A R S

The American Ceramic Society

100 YEARS

Published by The American Ceramic Society
in celebration of its Centennial 1898 – 1998

The American Ceramic Society is grateful to these Centennial Celebration Contributors,
without whose generosity this book would not have been possible:

BENEFACTORS

Corning Incorporated (see p. 164)

Kyocera • AVX Corporation (see p. 156)

PATRONS

Murata Manufacturing Co., Ltd. (see p. 151)

Ferro Corporation (see p. 145)

SPONSORS

Zircoa

Dow Chemical Company

3M Corporation

FRIEND

The Association of American Ceramic Component Manufacturers (AACCM)

Creative Development
Jane Mobley

Research
Linda Lakemacher

Design
Vivian Strand

Writing
Robert Butler
Samantha Adams • Meghan Humphreys

Editing
Jane Mobley • Linda Lakemacher • Michael DeMent • Lyn Foister

Production Art
Vivian Strand • William Petitt • Samantha Adams • Todd Myers

Research Assistance
Carman Stalker • Debbie Stalker • Brenda Horn • Mary Leonard

Printed in the United States of America
Richardson Printing, Inc.
Kansas City, Missouri

Library of Congress Catalog Card Number: 98-70444

ISBN: 1-888903-04-X

DEDICATION

This book belongs to the members of The American Ceramic Society,
past, present and yet to come.
For 100 years, this Society has found its strength in its members.
It is fundamentally a volunteer organization,
distinguished by people who care,
who are enthusiastic about their own work and the work of others,
and who remain committed to keeping up the connections
that give the Society its life.

To that volunteer spirit, this celebration is dedicated.

▲

CONTENTS

PREFACE

by Richard E. Tressler and David W. Johnson, Jr.

The American Ceramic Society was founded in Pittsburgh in 1898, with a core of founding members who saw the need for a technical forum for researchers, engineers and practitioners of the art of ceramic manufacturing to share their work and experiences.

That basic motivation remains today for members of the Society around the world. And it underpins in many ways the creation of this book, which was conceived by David Kingery and President Jim McCauley, and further developed by Past President Bill Payne, as a pictorial and anecdotal account of 100 years of interaction and sharing among members of the Society and of the companies and institutions they and the Society help shape. As co-chairmen of the Centennial Celebration for the Society, we are very pleased to have helped make the dream come alive and to present this handsome volume to each member of the Society.

At the 1995 Annual Meeting, (from left) immediate Past President Richard Tressler and Sue Tressler and President David W. Johnson, Jr. and Bonnie Johnson greet guests.

The Society, of course, represents a vast world of personal interest, intellectual inquiry and artistic and commercial application. And though we all too often encounter the belief framed by the question — "Ceramics, isn't that pottery?" — we understand that the answer is much more complicated than that for many of the more than 10,000 members of our Society.

For most of us, what we do is all about "inorganic, nonmetallic materials processed by the application of heat." This is a remarkably concise definition that highlights the enormous range of interests represented by The American Ceramic Society and its members. Its only shortcoming is that it defines it by the negative, by what it is not; the reality of the Society and its members' interests and relationships is so much more positive than that.

After all, most of us got into the field of ceramics because of what it could be, for the potential it offered. In our case, the route for both of us was similar. We knew we wanted to be in a technical science and engineering field. In the orientation process before undergraduate studies we found that ceramics offered a breadth of science and technology that was appealing and the beginning of a friendship; in fact, for the two of us, we were in undergraduate school together and in graduate school shared a lab/office and a thesis advisor.

So that's why we are here, but why are we so involved with ACerS? Why are there so many exceptional individuals deeply involved with and committed to the ongoing success of this Society? These are questions to ponder in our centennial year, and we'd like to offer at least some answers that seem to make sense to us:

Professional Identity

The Society offers us an opportunity to gain recognition for the important work any of us does in our chosen field. The recognition may be informal: a colleague's congratulations on a job well done. Or it may be a more public recognition offered in the guise of an organizational award for research, teaching, leadership or other achievement. Not only are these personally satisfying, but the respect given these awards and their recipients both inside and outside ACerS means they can play key roles in building a person's professional advancement through promotion or tenure.

Professional Development

ACerS provides us a forum for giving talks and publishing papers. For those of us in academia and industrial research, part of our output is what we say and what we write. We are judged by the talks we give and the papers we publish. It is important for us to have a prestigious society as this forum. In addition, Society meetings and publications are crucial sources of up-to-date information on what the latest research in our field is, thus providing us the information and the stimulation for new ideas in our own work.

Personal Networking

The Society keeps us in touch with others in the field. None of us works in the center of the technical universe. It is a place where we meet with others in the field and discuss what we are doing in an informal setting. Often what goes on in the halls is as important as what goes on in the meeting rooms. One of the wonderful things about ACerS is that it is a place where we can meet and talk to all the important people in our own interest areas.

The value of this can simply be social. We tend to become separated from those we have been close to in the past whether it be in our education or in previous jobs. ACerS gives us an opportunity to see those people again and spend some time with them. Our meetings all have some social element, which makes the Society one of the best people-support societies.

Beyond that, some of our most enduring friendships have formed through Society connections.

The value of this also can be professional. Especially for students, but for others as well, the Society is a place to meet those who may be future employers or a place to get your resume seen. It is simply the most effective way to find what the opportunities are in the field.

Professional Responsibility

All of us have a responsibility to support our profession and enhance its visibility and credibility if it is to grow along with our satisfaction in being a part of it.

That support can take many forms; we are supportive simply by being members and attending Society meetings and functions. Or it may require a more demanding investment of our time and energies as advocates for our profession and the Society. An important way to accomplish that is to seek out and take on leadership roles within ACerS.

Speaking for ourselves, we did not join the Society with the intention of being leaders. But as we worked on division programming, we were recognized as those who could lead divisions or classes. For us this led to Society offices and the presidency. As you can imagine, nomination for the president of ACerS causes one to do a lot of introspection. Am I up to the job? Will I be respected for the job I do? Do I have the time to devote to it? Will my employer be supportive?

In the end, each of us accepted, and we found it to be one of the most rewarding professional experiences in our lives — something that could be said about all aspects of our involvement with ACerS.

That's why we are proud to say — and hope you are, too — that The American Ceramic Society has been and will continue to be a crucial part of our professional lives. We feel fortunate to be part of it as we commemorate this great organization's 100th birthday, and we hope that this book will help all of us rediscover and celebrate our rich heritage — and bright future over the next 100 years. ▲

INTRODUCTION

by W.D. Kingery

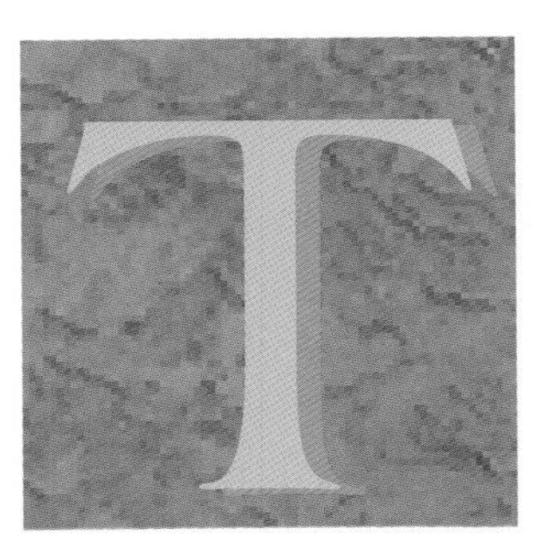

Those of us in The American Ceramic Society are truly fortunate. Ceramics is a wonderful field of activity. We are one with the Shamans of 26,000 years ago who shaped and fired animal and human figurines, and tested their mettle in the fire. We share and continue the work of potters from 10,000 years ago who discovered and used the plasticity of clay and the beneficial effects of fire to produce utilitarian household wares. We participate in the achievements of artisans who transform lowly clay into objects of outstanding beauty and value. We devote ourselves to the cutting edge of new technologies that are transforming the world in which we live. We can take both joy and pride in being members of the ceramic fellowship and part of this centennial celebration.

This sculpture was produced at Dolni Vestonice some 26,000 years ago.

This sculptural group done in glazed terra cotta by Luca della Robbia is in the Florence Cathedral.

In addition to the founders and key players who have shaped 100 years of The American Ceramic Society, there are other role models for those concerned with the transformation of ceramics from a craft to a science-based industry in the last 100 years. Among others, this list would have to include:

- Count Ehrenfried Walther von Tschirnhaus, who studied mathematics and physics at Leiden and used his "burning glass" solar furnace to carry out studies of Saxon minerals aimed at porcelain development;
- Johann Friedrich Böttger, a transformed alchemist who managed the laboratory program and announced success in achieving dense white European porcelain on March 28, 1709;
- Josiah Wedgwood, the most famous ceramist of the 18th century, who introduced into Europe the factory system of division of labor in concert with scientific experimentation that led to economic production of high-quality wares;
- Hermann August Seger, who, 100 years after Wedgwood, studied the chemistry of brick firing, investigated the physical properties of clay, introduced the idea of equivalent formulae for glazes and bodies (which are still used today) and also developed a series of compositions for pyrometric cones that are still used for determining a combination of time and temperature during firing. Seger's collected writings were translated and published by The American Ceramic Society as its first publication; and
- F.H. (Ted) Norton, who was trained in physics and came to the Massachusetts Institute of Technology, where he founded a graduate program in Ceramic Science that included thermodynamics, colloid chemistry creep deformation at high temperatures, all aspects of Refractories and a series of seminal researches on the properties of clay. Norton was an accomplished potter as well as a ceramic scientist. His vision is indicated by the fact that he had designed and built in his laboratory one of the earliest transmission electron microscopes used for ceramic studies. His integration of the chemistry, physics and geology of ceramic materials was spelled out in his 1953 text *Elements of Ceramics.*

Of course, there are others who deserve to be recognized as well. For example, the particle size of clays is very small and the resulting microstructure of clay bodies is difficult to visualize with the optical microscopy techniques of geologists. During the first half of the Society's existence, the instruments necessary for a true ceramic science were invented.

Among the key players in this effort were Max Von Laue, who observed

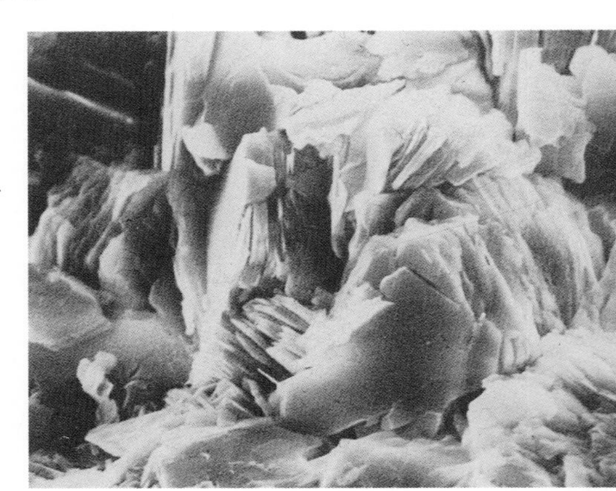

The individual particles of kaolin clay are fine platelets with a thickness less than half a millionth of a meter. Observation requires use of an electron microscope.

diffraction of X-rays by crystals in 1912. Already by 1915, W.L. and W.H. Bragg published the structure of crystals determined by X-ray spectroscopy. A year later, Debye and Scherrer developed methods of powder diffraction allowing the internal crystal structure of ceramics and glasses to be experimentally studied. About this same time the phase equilibrium rule of J. Willard Gibbs was being applied to difficult high-temperature silicate systems, mostly for purposes of geology but including systems essential to understanding ceramics. In the late 1930s the electron microscope was invented, but it only came into widespread realization after the end of World War II, at the close of the first half century of our Society's existence.

A decade or so after the end of World War II, the century-long task of transforming ceramics from craft tradition to science-based industry was pretty well complete. More than half of an "ideal" ceramics engineering curriculum consisted of basic science and engineering sciences. "Engineering Science" became the linchpin of university research. This changed the social structure of the university-ceramics industry relationship. University ceramic research came to be based on the maturing of various strands of the science of materials and the parallel exponential transformation of scientific instrumentation.

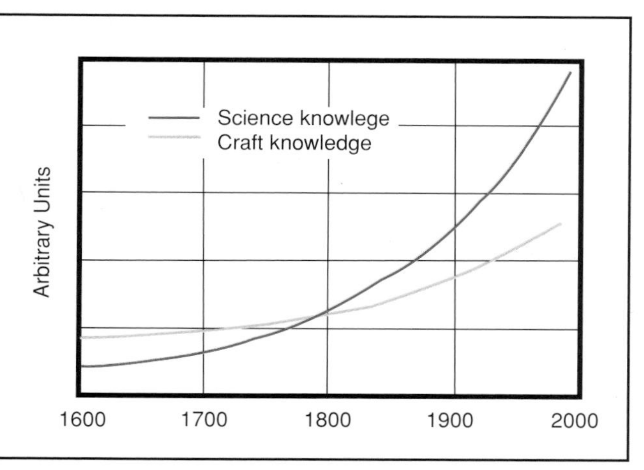

A very schematic illustration of the relative growth of science knowledge and craft knowledge in ceramics.

These changed relationships had an impact on the traditional ceramic industry in relating to the development of new science and new researches. In 1950 nearly all of the articles published in the *Journal of The American Ceramic Society* were industry related; by 1960 only a minority of articles were related to commercial product manufacture or use. The Basic Science Division of The American Ceramic Society was established in 1951, and its development mirrored some of the changing interests among ceramists.

Over these years the traditional ceramic industry had a decreasing role in controlling the direction of university researches. Science knowledge became further and further separated from clay manufacturing industry knowledge. An increasing number of The American Ceramic Society's members found their central professional focus was related to high-tech ceramics.

In the last decade or two a transformation of manufacturing technologies has occurred toward recognition that they are systems involving suppliers, distributors and customers on an equal basis with factory operation. At the same time the computer and information revolution has transformed the nature of the workplace and brought science and technology to be intertwined with activities on the factory floor. Science and practical application — the first debate of the Society at its formation a century ago — is still an important topic today.

Indeed, the Society serves both scientists and all those involved in manufacturing ceramics. We stand now at a new point in our development since both ceramic science and ceramic industry are enmeshed in a global environment. The days of local ceramics, regional ceramics, national ceramics — even continental ceramics — are ended.

It is a hopeful sign of the times that The American Ceramic Society is well on its way to becoming an international entity. Since the first trips abroad to meet ceramists in Europe, taken by Society members at the beginning of this century, the value of open exchange without boundaries has been obvious. Today, more than 20 percent of our members work and live outside the United States, and the numbers are growing. With modern communications and the Internet, members half a world away can easily share ideas, inquiries, and solutions.

Looking ahead to a new century, it's clear that our future successes will lie in extending the Society's global reach for the benefit of all its members. It's a pleasure to see how far we've come — and to imagine how far we yet can go. ▲

W.D. Kingery

Fiber optic cable, made of tiny strands of optical glass, is used in various capacities by industries such as telecommunications and medical technology.

A CENTURY OF DISCOVERY

The beginning of a second century is an important marker in the life of any organization. It is a high point that gives us a longer view of history than we usually stop to see; it invites us to stop and look both forward and back. • This long view lets us skim the details — What things happened when? — and get to the essentials that can only be revealed over time: Who are we? What holds us together? Where can we go from here? • Ceramics attract people who have a passion for useful — and beautiful— results. Turned toward association, this passion has provided the dedication to process that an organization must have to last. Over the course of 100 years, through the commitment of thousands of people, ceramists and their colleagues in allied industries have created an association so diverse in its interests and membership as nearly to defy definition. But the common element mixed through all of it is the spirited blend of expertise and camaraderie that The American Ceramic Society has offered its members since its beginning.

The world's largest crystal ball (left), in the lobby of The American Ceramic Society headquarters in Westerville, Ohio, and created by artist Christopher Ries from glass donated by Schott Glass Technologies, Inc., seems an emblem of the future. (Above) A century ago, few consumers knew that the decorative glass they prized was a ceramic product.

WHO ARE WE NOW?

The American Ceramic Society is among the most comprehensive of all technical and professional associations in the variety of its members' interests. Although ceramics is the common ground, the very nature of ceramics creates a world of opportunity for inquiry, creativity, production and distribution. That diversity makes neat descriptions of the members and their areas of expertise very difficult.

"A ceramist is anyone who does ceramics," said one long-time Society member wryly, "and anyone who does ceramics is a ceramist."

The founders of The American Ceramic Society came together to enjoy a mix of science and shop talk in a social setting. The science of ceramics, its trade and commerce and the fellowship of peers are the elements that have made up the organization from the start.

Categories abound in Society records, but they never seem to take in quite everyone and every interest. In preparing for the Centennial celebration, in a moment of spirited committee discussion, former Society President and Centennial Co-Chair Richard Tressler offered a model elegant in its simplicity.

"The Society is a vertical integration," he said. "What we do starts in science, where the possibilities are explored. Then it moves to creation of potential products. Then comes the development of the products — and all the raw materials, manufacturing, plant activities that are so much a part of ceramics are here. Then there are the sales and distribution of ceramic products, getting

The international headquarters of The American Ceramic Society in Westerville, Ohio, welcomes visitors to a building made as much as possible of ceramic materials.

The American Ceramic Society is one of the world's most comprehensive technical and professional organizations.

The Ionic Structure of Glass, *by Dominick Labino, is a mural depicting the basic structure of silica glass. It is displayed in the* Ross C. Purdy Museum of Ceramics.

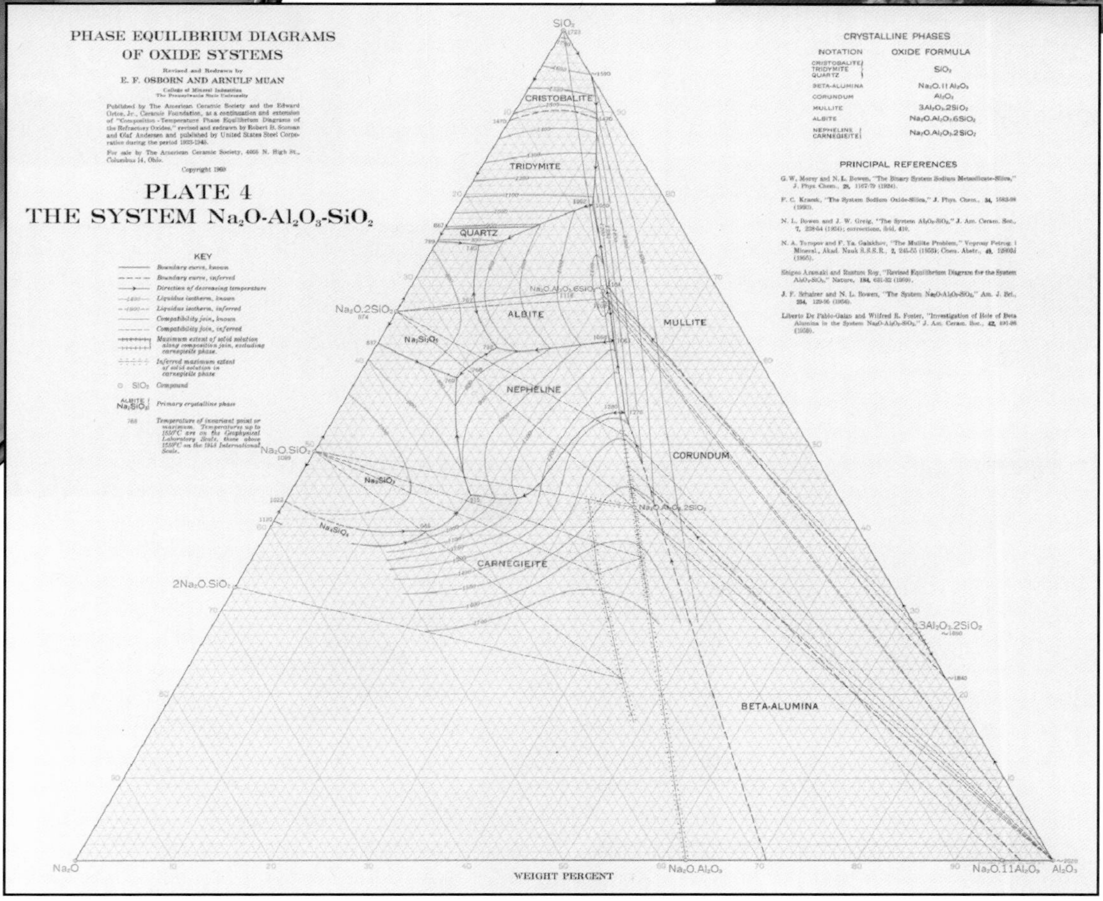

By many members' assessments, the two most important accomplishments by the Society in behalf of ceramics worldwide have been the publication of phase diagrams (left), a program begun in 1933, and the meetings (above, 1995 Annual Meeting) that encourage fellowship and the exchange of ideas.

them out into people's lives. Who we are in the Society is all of those."

From science to products people use. From ideas, through discovery, to design and development, manufacture and distribution into everyday life — that wide range is The American Ceramic Society. Its members all over the world represent the spectrum of those fields of interest.

They bring to the Society their differences — in background, interests, opinions, direction — as well as their shared intent to create an association. And so the first century has been spent finding rewarding ways to work (and play) together. Like any good story, this hundred years' tale has its ups and downs, moments of disappointment and triumph, memorable characters and a cast of thousands of outstanding people who have worked diligently without fanfare.

Is there one fundamental thing that might be said to characterize the Society in its one hundredth year? Any generalization about such an eclectic, yes *ceramic*, group, is probably unfair. But what is the Society essentially about?

"It's about love," exclaimed Jim McCauley, president during the Society's 100th year. "That's what the Society is really about — love for this profession. We love what we do. We love our work. We love what the Society stands for in science and industry and education and art. And mostly — except for the inevitable small squabble here and there — we love being in it together."

WHAT HOLDS US TOGETHER?

The Society's diversity is a celebration of the diversity of ceramics itself as a field of study and achievement. People are drawn to work in ceramics for reasons as myriad as the properties and possibilities of the materials and processes themselves. And they are drawn to the Society to share what they know and hope to learn about work that belongs to

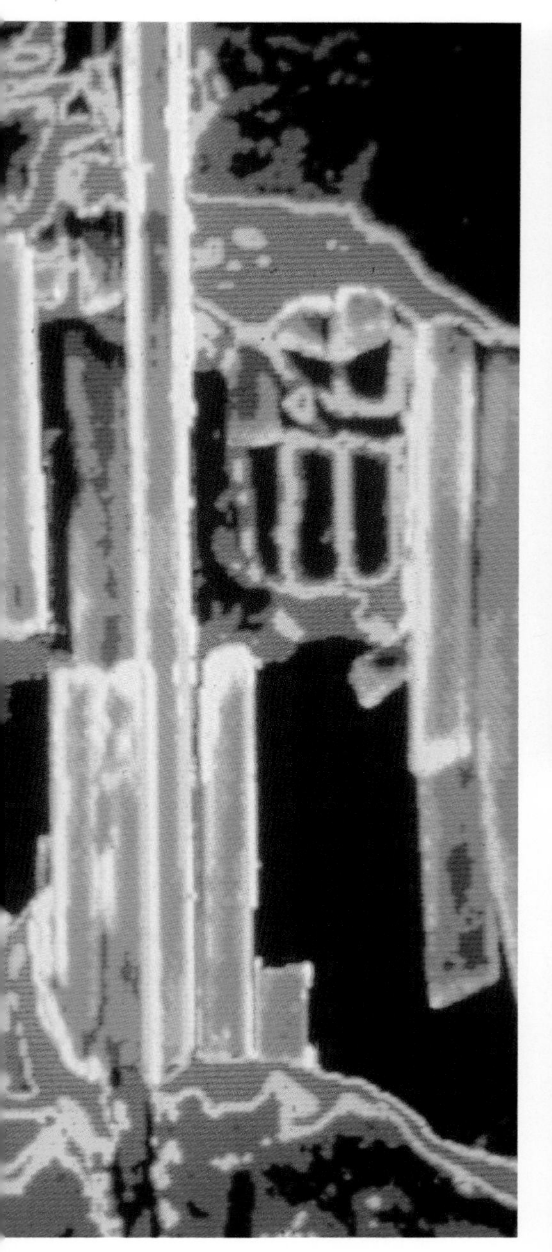

(Clockwise from lower left) Biomedical ceramics like this hip joint are a promising field for the 21st century.

Precise calculations by ceramists and by modern tools such as the scanning electron micrograph are behind many of the products people take for granted.

Outgoing President Carol Jantzen passes the gavel to new President James McCauley at the 99th annual meeting.

Factories like these were represented by brick men at the convention where The American Ceramic Society was first imagined.

Trips by Society members to visit English ceramists early in the century set a tone for international exchange.

the dawn of humankind on Earth and still holds promise to reach the stars.

Ceramics are essential to human life, and are among the oldest, the largest and the most essential of all of man's endeavors.

The Earth itself may be pictured as an enormous ceramic ball, about 8,000 miles in diameter and 25,000 miles in circumference. It began as a huge molten mass, a seething heterogeneous mixture of silica, mica, alumina, feldspar, rutile and other igneous rock minerals, compounds and elements of the periodic table. Some 10 billion years ago, this mass began to cool and solidify, then developed an atmosphere to encourage and sustain life as we know it.

When we consider that the oxygen, silicon and aluminum so important to ceramics form approximately 85 percent, by weight, of the elements of the Earth's crust, we can form some concept of the magnitude of the foundation upon which the ceramic industry stands. The discovery and use of ceramics were fundamental as human beings developed, found the value and applications of tools and began to create societies.

Ceramics are so numerous and so much a part of the way people inhabit the Earth that they are taken for granted. Every country of the world, even the most underdeveloped, has its version of a ceramic industry.

What are ceramics, really? Even the earliest technical literature concerned with ceramics finds the authors struggling to define just what they mean by the word.

Organizers of The American Ceramic Society declared that "clay, glass, cement, enameling, mortar material and refractory material industries and others of an allied nature" were to be included in the field of ceramics. Meanwhile, the British equivalent of the Society thought "ceramics" should refer only to pottery,

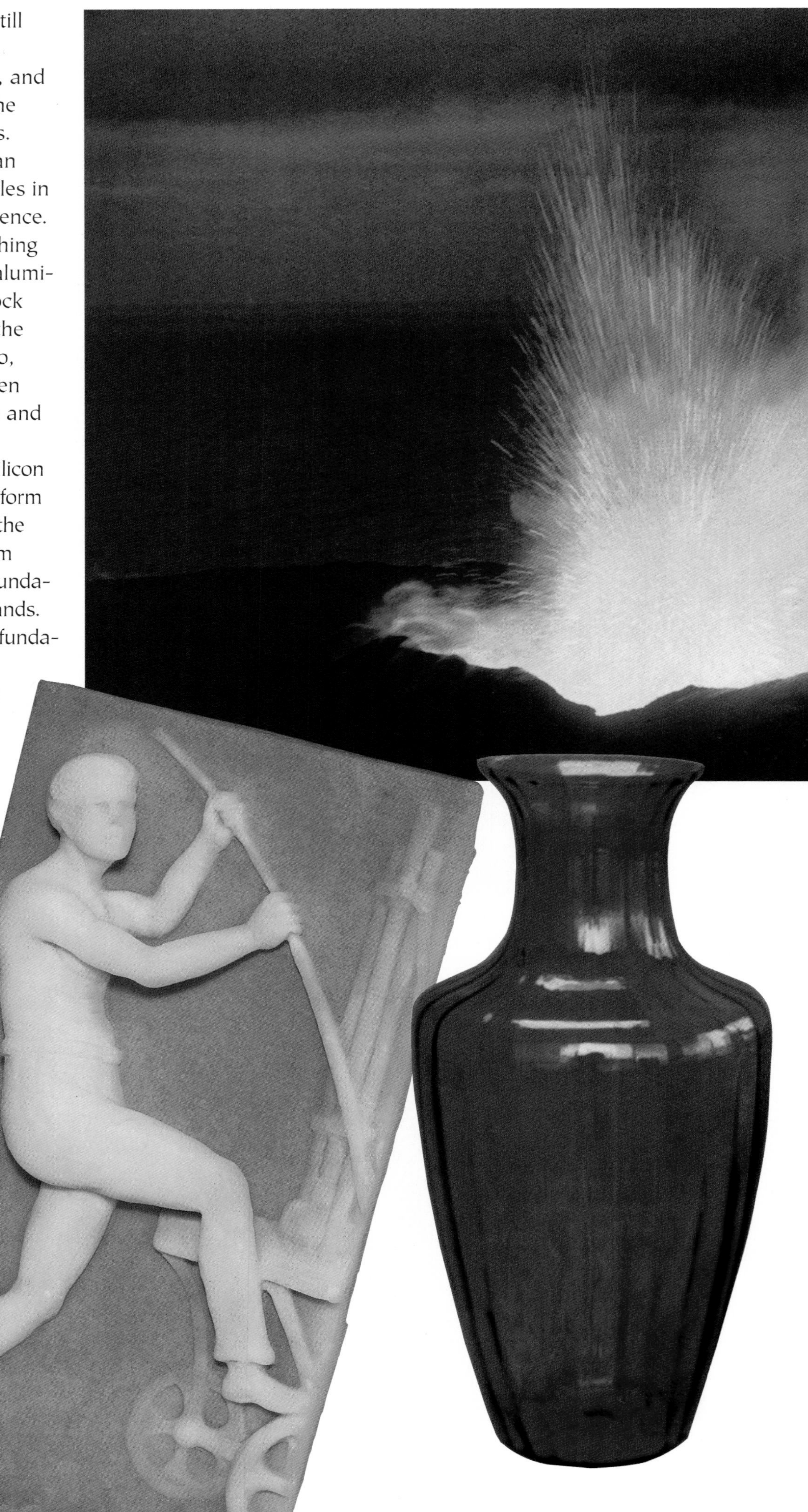

Ceramics belong to the dawn of humankind on Earth and still hold promise to reach the stars.

(Clockwise from lower center)

An example of "industrial artware," this 1931 vase was designed by Frank Ferrell and made at the Roseville Pottery Company.

Tile is one of the oldest and still today the most popular form of ceramic products in daily life.

From the molten center of the earth to sophisticated kilns in industry, heat is the common element fundamental to ceramics of almost every kind.

The Star Chart *sculpture in Society headquarters is a montage of ceramic shapes.*

In the 1930s, in the Pisgah Forest Pottery, hand work was still a primary means of ceramic production.

not to brick and refractory materials.

According to a committee assigned to get to the bottom of the issue, the word *ceramic* wasn't used in the English language to any significant degree before the mid-19th century. An 1850 edition of J. Marryat's *Pottery and Porcelain Introduction* refers to "the plastic or keramic art." By the 1868 edition, the spelling had been changed to "ceramic." Nuttall's *Dictionary of Scientific Terms* (1878) defined "ceramic" as "denoting the plastic arts; a term frequently applied to ornamental pottery."

At the request of the Society, William A. Oldfather, a professor of classics at the University of Illinois, did some digging into the word's origins. Ceramic appears to have been derived from the Greek word *keramos*, which is related to some primitive Indo-European root that means "burn" — perhaps the Sanskrit *car-* or *cra-* which meant "cook."

Sages in ancient India used to carry earthen pots filled with water. These pots were called *kro*, which is derived from the word *kra*, meaning baking or frying. It's also been noted that *keramos* seems a close cousin to the Latin *cremare* or "to burn."

Over time, various translations led to disagreements over exactly what the word *ceramic* meant. Some felt it referred to clay. Others held that it referred to pottery. Some stated unequivocally that the word should be used only to describe a vase or urn. In Greek, one who produces *keramos* was called a "kerameus," the plural of which was "kerameis." The district of ancient Athens near the Dipylon gate was known as Kerameikos, or the "quarter of the potter."

Oldfather attempted to give the broadest possible interpretation: "The Greeks (probably)

Before recorded history, people made earthenware vessels like these, and today archeologists depend on ceramic objects to help date cultures. In 1951, examination of artifacts from an excavation in southern Ohio suggested that a small ceramic and metallurgical industry might have existed there as early as 500 years before Columbus landed on San Salvador.

If this is true, the ceramic industry predated the founding of the United States by 800 years. Today high-tech ceramics, such as Zerodur glass ceramic, meet the quality standards required for the world's largest glass monolith (above), used as a mirror carrier in giant telescopes that help look skyward for the possibility of new worlds.

At the turn of the century, the ceramics industry employed women in production, as in a Pennsylvania factory (above) and Denmark's Royal Copenhagen Porcelain Manufactory (left), and talented painters were especially prized. Today, when technical solutions abound, the hand-painted quality of work such as Royal Copenhagen's Blue Fluted pattern, with more than 1,000 brush strokes per plate, is still appreciated.

had occasion to use *keramos,* or some derivative to denote the complex of allied and derived industries, because of the meaning of the stem and the primary place of the potter's art, both generically and in relative importance."

In general terms (and the problem with categorizing ceramics is that there always seems to be an exception to the rule), a ceramic is any inorganic, non-metallic material or product subjected to a temperature of 1,000 degrees F or above during its production or use.

But even that definition is in transition. Research over the past 30 years has developed thousands of new ceramic applications; some people have suggested these new procedures and products require new definitions of the very word *ceramics*. For centuries, *ceramics* has been almost synonymous with "clay." But many of today's new ceramics contain little or no clay. For that reason it has been argued that ceramics be redefined as having their origins in any inorganic substance except a metal. Trouble is, there's an exception even to that rule. Carbon is by most definitions not a ceramic material, yet it is used as a ceramic because of its highly refractory nature.

It is a testament to the marvelous versatility of ceramics that even the term continues to be subject to steady reinterpretation and new definitions. Let us not define ceramics by what goes into them, but rather by the process that creates them. By this definition, a ceramic is an inorganic material that at some point in its manufacture undergoes the application of heat to render it hard and resistant.

Most ceramic products are shaped at room temperature, then permanently hardened in a firing process. Originally they were made of clay or some other natural earth containing silica; today they may be fashioned from a huge menu of elements. Most ceramics consist of complex oxides and silicates, though boride, carbide and nitride ceramics are also enjoying widespread use. New materials are fashioned from high-purity materials formed and fired by innovative techniques, and they are usually developed for the electronic, nuclear and aerospace industries.

Ceramics belong to many industries from mining to the decorative arts. A strip chart recorder about 1955 (left) appears almost as primitive by today's standards as a brick mill of the 1880s (above right). The steel industry's need for heat-resistant materials was a major stimulus to ceramic science and technology (above left); heat is equally important to the glass maker's art (below), exemplified by these Labino paperweights in the Ross C. Purdy Museum of Ceramics.

Ceramics are so much a part of the way people inhabit the Earth that they are taken for granted.

More than 70 million tons of raw materials are processed by the United States ceramic industry every year, ranging from clay that is, literally, dirt cheap, to rarer substances that must undergo extensive processing before they can be used.

Mica is a good example of a naturally occurring ceramic. This common mineral is often used as electrical insulation. But most ceramics are produced by mixing raw materials, then subjecting this blend to intense heat to create new mineral forms.

Much more so than metals and organics, ceramics can be tailored for particular uses. A skilled ceramist can turn out a ceramic product light enough to float on water, or with an adjustment of materials and processes, he can create a ceramic denser than lead.

Ceramics are incredibly strong and stable. They are particularly resistant to high temperatures and largely impervious to hot corrosive gases and molten metals and to most acids.

Only a small percentage of the ceramic products created today are consumer-oriented. Most are made for industry with specialized applications in nuclear reactors, space vehicles, pumps, valves, metal processing furnaces, computers, protective coatings and optical equipment. Taken as a whole, the ceramic industry is so vast and complex that it boggles the mind.

It includes the development of methods and machines for mining silicate earth, moving it and processing it, for the creation of crushing machines to reduce large pieces of rock to small particles, and screening machines capable of separating different-sized particles.

Other machines mix the various materials required for ceramic products. Mills blend ceramic materials and water to the desired consistency, while other machines form, press, mold, cast and extrude the material into the required shape.

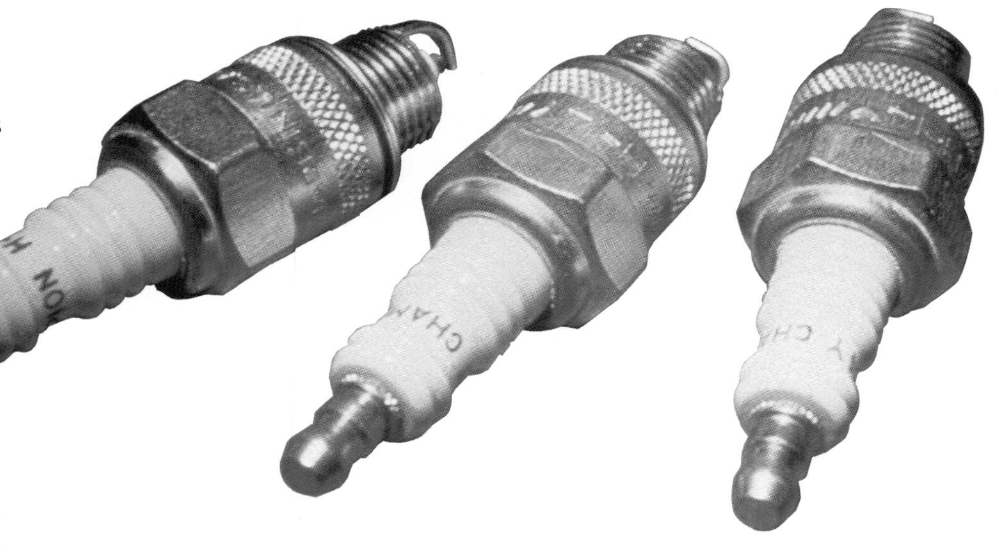

There are machines to grind, apply coatings and otherwise finish ceramic products to meet a wide variety of customer specifications. In the case of some technical ceramic parts, these objects

Ceramics come in all shapes and serve an almost indescribable range of functions. (clockwise from lower left) From the spark plugs in the first "horseless carriages" to today's high performance cars, ceramics are important to the automotive industry.

A view of Cincinnati's Albert B. Sabin Convention Center, where ACerS annual meetings have been held periodically in past years, highlights the varied use of ceramics in urban spaces.

From golf cleats to cookware that withstands both heat and cold, ceramics find uses in daily life — as well as on the extraordinary journey, such as the one made by this tile built for space shuttle heat control.

Machines that make ceramic shapes, such as this mullite extrusion in a factory in the 1940s, have made up a crucial part of the ceramic industry.

can only be measured at specific temperatures and humidities, and the machines that allow the creation of these conditions must also be the concern of the ceramist.

Kilns, furnaces and drying machines control the temperature, humidity and air flow around ceramic products, which otherwise would be subject to cracking under unmonitored conditions.

Moreover, sophisticated tests have been developed to analyze the chemical content of clays and other silicate soils; equipment has been created that can test ceramic products for strength, hardness, elasticity, thermal conductivity and expansion, electrical conductivity and a host of other critical properties.

And because the ceramic industry relies on intense heat to effect a coalescence of ceramic powders, it is intimately concerned with those associated industries that provide fuel for this process. These include the natural gas, fuel oil, liquid petroleum gases and coal industries.

From the cool precision of ceramic templates (below) or the honeycombed surface of a catalyst support (center) to the refractories that make possible high temperature furnaces (above), ceramics have been an integral part of modern industry, in great factories and in smaller concerns like this kiln brick operation in 1955 (lower right). But for many people, "ceramic" conjures the elemental shape of a vase, like this one (upper right) made in the 1920s by the Weller Pottery Company, painted and signed by Hester Pillsbury, one of the firm's most talented and prolific hand painters.

CERAMICS ALL AROUND US

It is almost impossible to open one's eyes without seeing a ceramic product or a product that depends on the ceramic engineer or the ceramic scientist for its existence. When people in general talk about ceramics, they usually means dishes, pottery and fine art figurines. And those are certainly ceramic products. But they barely scratch the surface of today's ceramic industry.

Some products are obviously ceramic in origin: bricks, tile, tableware, flower pots, lavatory fixtures.

But glass, in all its thousands of permutations, is also a ceramic product, from the spectacles on your nose to the windows of skyscrapers to the glass fibers in the drapery and the fiber-optic cable that brings the picture to your television.

Mythology lends its own colorful interpretation to how "ceramists" came to be.

One account revolves around the Kerameis — the potters of ancient Greece — who sacrificed to the god Keramus, son of Bacchus, the god of drinkers and infamous for the Bacchanalian orgies practiced in his name.

Their name came to be synonymous with ceramics, by some accounts, when the ancient deities were each taking under their protection different trades and professions. Diana became the goddess of hunters, for example, and Neptune the patron of sailors.

According to this tale, toward the end of the distribution only two gods were left — Keramus and the Devil. The only two trades left unprovided for were potters and lawyers. Obviously, neither group of professionals wanted the Devil as a deity. To ensure fairness, a potter and a lawyer were blindfolded and placed in a room with the two remaining gods. The potter found Keramus first, and claimed him as the potters' tutelary deity. The lawyers, of course, were stuck with the Devil.

Ceramics have electrical applications as high-voltage insulators, in resistors and capacitors, as the memory in computers and as spark plugs in internal combustion engines, and more recently, in high-temperature superconductor applications.

Abrasives of all sorts are ceramic in origin, as are plasters and cements.

Heat resistance is one of a ceramic material's most attractive features, which is why ceramic tiles provide the heat shield for today's space shuttles. An entire class of heat-resistant ceramics called refractories make possible the construction of the kilns and blast furnaces and nuclear power plants that are the heart of modern industry.

Ceramics are all about us — in the cars we drive and the buildings we live in and in the sidewalks under our feet. They're even used by dentists as filling material and for the crowns and caps on our teeth.

Ceramic filters made from porous porcelain can isolate microbes and bacteria from milk and drinking water, separate dust from gases and remove solid particles from liquids.

Ceramics are essential to the construction industry, to the generation of electrical power, to modern communications, space exploration, medicine, sanitation — just about every field of endeavor. Ceramic semiconductors made possible the solid-state circuitry of transistor radios and portable TV sets that showed people on the move how to take their information with them and gave rise to a whole new way of thinking about education and entertainment.

Ceramic armor, which is both lightweight and resistant to impact, has been fashioned to protect aircraft, military vehicles and individual soldiers. Individual electronic components and complex multi-component integrated circuits have been fashioned from ceramics. Single-crystal ceramics have important mechanical, electrical and optical applications. Ceramics include items so delicate that they can be shattered with a strong tap, so tough that they protect our own fragile selves and so enduring that they remain after thousands of years to reveal to us glimpses of the ancient peoples who first fashioned them. ▲

Expositions provide a chance for the next generation of ceramists and ceramic consumers to get a glimpse of the industry. Today The American Ceramic Society is developing its educational programming to reach younger students with the wonders of ceramics.

The Earth itself is a huge ceramic ball.

Ceramics make it possible to retrieve astronauts from space, shielding the occupants of a spacecraft from the intolerable heat of re-entry into the Earth's atmosphere. Such applications were unimagined by workers a century ago, posing amid firebrick for kilns and furnaces. Who can guess what extensions of human imagination may take ceramic form in the century to come?

National Institute of
Ceramic Engineers

Brick men setting out for the 1898 convention of the National Brick Manufacturers' Association had no reason to think that year's event was going to be extraordinary. They didn't expect much difference from every meeting in the previous 11 years, and most of them were not looking for change. • They liked what they had in store — a holiday that relieved the year's bleakest season with the promise of three days in Pittsburgh's luxurious Monongahela House.

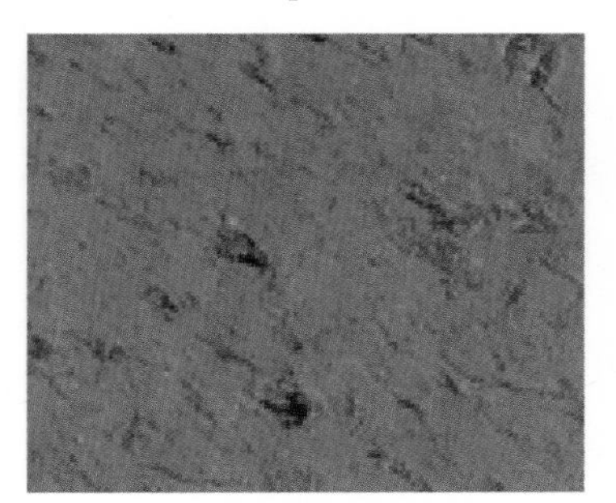

They knew they'd find plenty of good company, good food, brandy and cigars, and brick discussed from every angle. They planned to trade tips, indulge in conviviality and present papers with titles such as "The Use of Electric Tramways in Hauling Clay." • But among the practical, meat-and-potatoes topics, one presentation stood out. On February 16 a young man named Elmer E. Gorton, representing the American Terra Cotta & Ceramic Co. of Terra Cotta, Illinois, addressed the gathered conventioneers to deliver his paper, "Experimental Work, Wise and Otherwise." • It was change in the making, although most of the brick men didn't recognize it at the time. The substance of the paper was deceptively simple. It was the energy of the question it raised that would mark the ceramic industry for the next 100 years.

THE AMERICAN CERAMIC SOCIETY

(Above) Ceramic shapes form Star Chart, *a welcoming frieze in headquarters' lobby. It has been displayed in two of the Society's headquarters buildings since it was commissioned by the Society in 1986. The artist called it* Star Chart *for its use of light-colored pieces originally set against a dark blue background in the Brooksedge home of The American Ceramic Society.*

(Opposite) The American Ceramic Society's international headquarters building in Westerville, Ohio, was dedicated in 1991.

In his paper, Elmer Gorton described his efforts to develop a glaze for terra cotta. The topic wasn't out of line with the sort of thing the Brick Manufacturers were hearing from other presenters, but Gorton's approach was unique. As he described his glaze experiments, he filled a blackboard with chemical formulas and mathematical calculations, showing how he arrived at his conclusions. He also displayed trial pieces — terra cotta tiles with a variety of glazes showing how modifications in the formula affected the final product.

It was revolutionary. But most of the men in the room didn't know it. The delegates were polite but not terribly taken by Gorton's presentation. They were mostly "rule-of-thumb" men, who had learned the business by doing it. The equivalent of cooks whose recipes involve "a bit of this and a bit of that," they had little interest in fancy formulas and calculations.

Indeed, many of them possessed their own closely guarded recipes. They were not inclined to share these secrets with the competition, not even for the sake of science. One chronicler of that Brick Manufacturers' gathering noted: "It was apparent that an extensive addition of this sort to the programming of the Association would be unwelcome, on account of the comparatively small proportion of the members who could understand or follow them."

Gorton's presentation may have bombed with most of the delegates, but it made history nonetheless, for it was the first paper on a chemical topic ever presented to the National Brick Manufacturers' Association. And, with a little help from the weather, it became the spark for the creation of The American Ceramic Society.

Later that night Gorton sat down with Samuel Geijsbeek, a chemist with Sherwood Bros. and Co., a refractories plant in New Brighton, Pennsylvania, another young man who had been very interested in the notations Gorton had put on the blackboard. Gorton and Geijsbeek agreed that a strictly scientific approach to ceramics was the wave of the future.

Together they proposed to another convention delegate, the man who had taught both of them

CLASS A. XXc. No. 42044

Library of Congress,

Be it remembered, to wit:

That on the nineteenth day of September, 1902, Edward Orton Jr. Columbus, Ohio, hath deposited in this Office the title of a BOOK, the title of which is in the following words, to wit:

Transactions of the American Ceramic Society, containing the papers and discussions of the fourth annual meeting, held at Cleveland, Ohio, Feby 10-12, 1902.

the right whereof he claims as ~~author and~~ proprietor in conformity with the laws of the United States respecting Copyrights.

Office of the Register of Copyrights,
Washington, D. C.

Herbert Putnam
Librarian of Congress.

By Thorvald Solberg

Not only was Pittsburgh the city in which The American Ceramic Society was founded, but it has enjoyed a long association with the Society and with the larger ceramic industry.

More than a fifth of the Society's charter members hailed from Pittsburgh or later were connected there. In part this is because that region of Pennsylvania has always had ample supplies of clay and sand; even more important is the city's role as a steelmaking center.

In 1958 the Society dedicated a special plaque on the site of the old Monongahela House, pictured here in 1900, in recognition of the Society's creation. Made of a photosensitive glass developed by S.D. Stookey of the Corning Glass Works, the plaque reads, in part:

> *In the Monongahela House which occupied this site, The American Ceramic Society was conceived and its organization begun on Feb. 18, 1898 . . . to those far-sighted men we respectfully dedicate this plaque.*

(Left) By 1902, the Society was in the publications business with Transactions, *a volume of scientific papers, deposited in the Library of Congress.*

at The Ohio State University — Edward Orton, Jr. — to create a society "composed only of ceramic chemists or those who would understand ceramic work from the scientific side, and would be willing to share their information with their fellow members."

Their fledgling organization might never have gotten past preliminary discussions if not for a winter storm that dropped temperatures to the low teens and first covered Pittsburgh in snow, then soaked it in a downpour. With sightseeing out of the question, the forced confinement in the Monongahela gave Gorton, Geijsbeek and Orton plenty of time to think about exactly what they were proposing.

It made sense for their new organization to remain a part of the N.B.M.A. — after all, at the turn of the century most ceramic endeavors were connected with brick plants.

The first step was to canvass the N.B.M.A.'s members while they were all trapped in the hotel. An additional six men were identified whose scientific backgrounds would make them eligible for membership in the new group. The most noteworthy of these may have been Albert V. Bleininger, a ceramics student at The Ohio State University who later enjoyed a prestigious scientific career in Pittsburgh.

Others were Willard D. Richardson of Shawnee, Ohio; Ellis Lovejoy of Union Furnace, Ohio; William D. Gates of Chicago; Carl Giessen of Canton, Ohio; and Gustav J. Holl of Cleveland.

On the last day of the convention these nine met, named Orton as provisional secretary, and by sharing the names of other contacts in the ceramic industry, came up with another 15 or so qualified men who might be interested in joining.

They also determined that "membership shall be selected from those who are actively engaged in technical ceramic operations and who are qualified by education and experience to contribute to the advancement of the Ceramic Arts."

It was agreed that another meeting would be held a year from that date, to coincide with the regular N.B.M.A.

THE ORIGINAL NINE

These are the nine members of the National Brick Manufacturer's Association who on February 18, 1898, in the Monongahela House in Pittsburgh, formed the nucleus of what a year later would become The American Ceramic Society:

Elmer E. Gorton, chemist, American Terra Cotta & Ceramic Co., Chicago.
Samuel Geijsbeek, chemist, Sherwood Bros. and Co., New Brighton, Pennsylvania.
Edward Orton, Jr., The Ohio State University, Columbus.
Albert Victor Bleininger, The Ohio State University appointee to the N.B.M.A. Scholarship, 1897-98.
Willard D. Richardson, general manager, Ohio Mining and Manufacturing Company, Shawnee, Ohio.
Ellis Lovejoy, superintendent, Columbus Brick and Terra Cotta Company, Union Furnace, Ohio.
William D. Gates, president, American Terra Cotta & Ceramic Co., Chicago.
Carl Giessen, Royal Brick Company, Canton, Ohio.
Gustav J. Holl, manager, Ohio Ceramic Engineering Company, Cleveland.

convention. Early statements by Orton and others stressed that the proposed society "was not meant to be in any way antagonistic to the older organization," but that a scientific group was needed because "the nature of the organization [meaning the N.B.M.A.] absolutely precludes its ever becoming in any degree a scientific organization; if it did become such a one, it would at once cease to do the greatest good."

Those working to create this new organization agreed that ". . . the subjects which we wish to discuss are those connected with the application of science to the practical needs of the clay industry. . . chemical technology, the physics of drying and burning, the mechanics of manufacturing and in short all and any branches of technology touching this great industry will find welcome admission to our program."

Persons responding favorably to the idea were encouraged to come to the next year's N.B.M.A. meeting "prepared with a paper or note or some specific item of interest as his contribution to our first meeting."

The organizational meeting of a society for the "Advancement of Ceramic Technology" was held February 6, 1899, in Orton Hall on The Ohio State University campus, a day before the beginning of the larger N.B.M.A. convention at the nearby Columbus Southern Hotel.

Fifteen men attended the session in the geology department's lecture room. The furious correspondence launched by Orton during his year as provisional secretary paid off — the meeting was well organized, the agenda clear and concise and the membership ready to throw themselves into this new venture, even if a few were dubious about the Society's long-term prospects. This spirit of conservatism and doubt as to the value of a technical study of ceramic subjects would be an ever-present obstacle in the early years of the Society. Nonetheless, as would continue to be true, an equal spirit of enthusi-

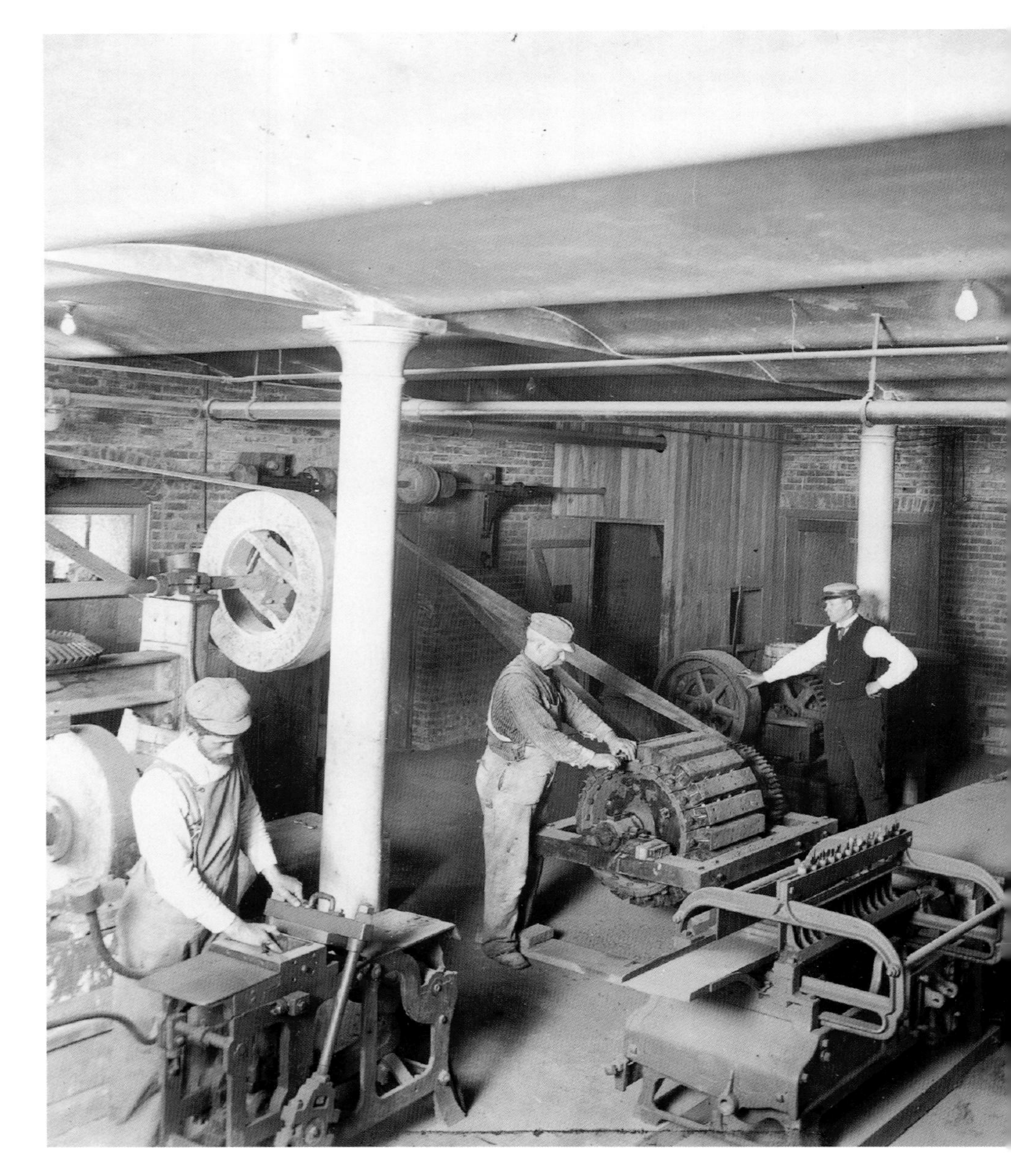

A selection of old programs from early get-togethers includes the first invitation sent out by Edward Orton, Jr. for a meeting on Des Artes Factiles, which soon became The American Ceramic Society.

A basement served as a work area at The Ohio State University, most likely Lord Hall, about 1900, roughly the time that the Society was getting under way for Edward Orton, Jr. (right, with watch chain and eye shade), who was serving as secretary of the Society and teaching in the ceramics program he founded at OSU. Ross C. Purdy (left, in a cap), who eventually became the Society's first paid general secretary, is shown operating some sort of brick press, in front of crushing and grinding equipment used for pulverizing dry clay. Behind Orton is a pug mill at the end of the table. The man in the middle is probably "Dad" Lysatt, department technician.

Edward Orton, Jr. was a founder of the Society and its most powerful figure for more than 30 years. Orton's remarkable vision for the Society is still a guide 100 years later.

asm for the possibilities balanced the nay-sayers, and the Society's organization moved ahead.

The first name proposed by Orton for the group was the Association of Ceramic Technologists. But by the next meeting on July 4, 1899, the group had settled on The American Ceramic Society.

William D. Gates of Chicago was named provisional chairman. A constitution drawn up by H.A. Wheeler, Stanley G. Burt and Elmer E. Gorton was discussed and adopted. After that the Society elected its first officer: Edward Orton, Jr. was given another term as general secretary. In fact, the tireless Orton would hold that post — without pay — for 18 years, until he reported for military service in 1917. During those years the Society's headquarters were Orton's office at The Ohio State University — or wherever this remarkable man traveled.

Other officers elected included H.A. Wheeler, president; Ernest Mayer, vice president; Stanley G. Burt, treasurer; William H. Zimmer, senior manager; Samuel Geijsbeek, second manager, and Charles F. Binns, junior manager.

The constitution adopted by those attending set out the exacting requirements for membership and additionally established an associate membership for those who could not yet meet those standards. Corporate memberships were briefly considered but were not instituted until several years later, largely because of fears by many members that their scientific Society might be overtaken by the participation of commercial interests.

Society organizers were determined that theirs be a scientific, not a commercial organization, one emphasizing the free exchange of ideas and research. "The status of its members in the industry commercially shall not be of importance in determining their attendance to the Association," read an early statement of intentions. "Their interest in the work and their attainments as ceramic technologists only will be considered."

It is worth noting that the original constitution wasn't amended until 1911. It was a set of rules that governed unchanged for a decade and with certain modifications was in effect for a quarter of a century — a testimony to the vision and pragmatism of the Society's charter members.

A second important action — and one that set a precedent for important publications from the Society — was approval of a proposal to translate into English the writings of Hermann A. Seger, the world's pioneer scientific ceramist. This project resulted in the publication of *The Collected Writings of Hermann Seger* by the Society two years later. Additionally, the membership authorized a study by the Committee on Equivalent Weights, which the following year would be published as the *Manual of Ceramic Calculations*, edited by Charles Fergus Binns.

These decisions secured the Society's position as a leader in the dissemination of scientific information in the ceramic field.

At this meeting, members of the Society also endorsed the creation of a Federal Bureau of Mines and authorized the drafting of letters to members of the U.S. Congress and the U.S. Geological Surveys urging the creation of such an agency — thus establishing the Society as a group willing to take an activist role when it came to public policy regarding the ceramic industry.

Before taking the chair, President Wheeler put in nomination the name of Mrs. Bellamy Storer (nee Miss Maria Longworth) of Cincinnati, as the first Honorary Member of the Society, on the grounds that as the founder of the famed Rookwood Pottery she was the first to produce a distinctive type of American art pottery ware. The motion was unanimously approved.

In his keynote address, Secretary Orton set out an agenda that in many respects flew in the face of the ceramic industry at that time. Among other things, he declared that the secretiveness which in the past had characterized and been the stumbling block of the ceramic industry should give way to liberal, scientific discussions; that the potter who had learned his mixtures and glazes as confidential heirlooms should be educated to appreciate that he would gain, rather than lose, by broad-minded and open discussions; and that the meetings and exchange of experiences of many technical minds meant progress.

It was a noble ambition and enormously appealing to Orton's audience that day, but it was heretical to the old-style ceramic technicians represented by most of the brick men who had been in Pittsburgh the year before. From the beginning, the stage was set for ongoing debate over science and pragmatism among ceramists in general and Society members in particular.

Recalled Orton of the first Ohio meeting: "The technical session to which we adjourned that day set a pattern of high scientific excellence that has been carefully emulated in the Society's annual meetings and publications in the years that followed."

During the formative years, the volume of papers presented at the Society's meetings was so heavy that in 1904 *Transactions of the American Ceramic Society*, the Society's first periodical collection of technical presentations, began publishing papers in advance. This was intended to permit a more comprehensive

In the early 1900s, clay products were fired in kilns like this one at Mason Color in East Liverpool, Ohio.

MEMBERS AND ASSOCIATES

The constitution adopted by original members of The American Ceramic Society required that full members possess technical ability:

> *We require that a man who aspires to place his name on the roll of membership shall prove by some specific contribution on the floor of the Society this fact of his ability to sustain himself there. The second requirement is that technical ability shall be supplemented by a desire to impart of his stores of knowledge and to serve his neighbor and mankind.*
>
> *Recognizing that the high standards required for full membership in the Society would preclude the participation of many persons intimately involved in the ceramic field, though not necessarily in the scientific or technical arena, the charter members voted to create associate memberships. These would be made available to those 'interested in ceramics and the allied arts.'*

Indeed, over the years thousands of persons came to full membership in the Society only after an "apprenticeship" as an associate.

Associates were entitled to all the privileges of membership, except that of holding office and voting. Those were reserved for full members.

Today the grade of full membership embraces all those more than 26 years of age who are interested in the field of ceramics. The grade of associate member has been modified to encourage participation by those younger than 26, who do not qualify as student members, by allowing them to join at a reduced rate.

From the start, Society meetings promised almost as much joviality as science. From the first division meetings, golf was a regular part of the program for many members.

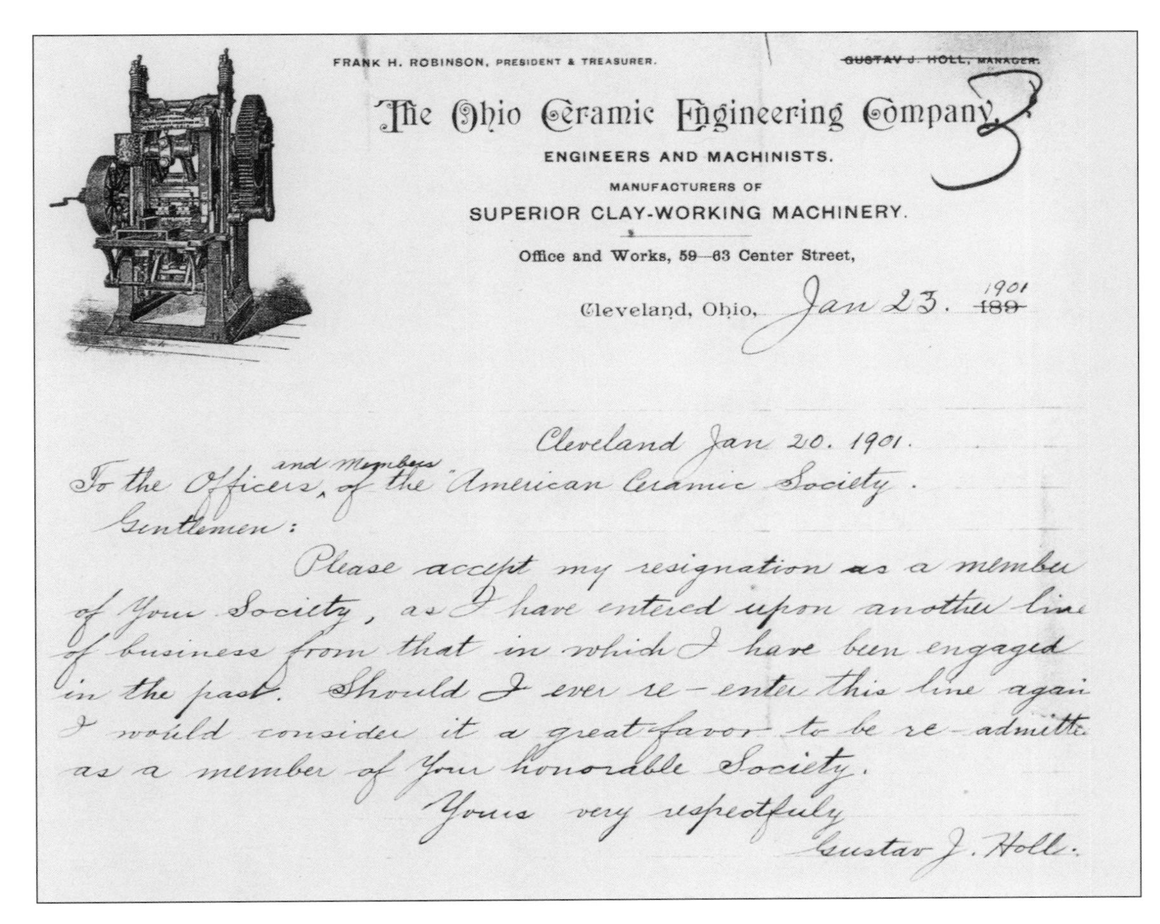

FRANK H. ROBINSON, PRESIDENT & TREASURER. ~~GUSTAV J. HOLL, MANAGER.~~

The Ohio Ceramic Engineering Company.

ENGINEERS AND MACHINISTS.

MANUFACTURERS OF

SUPERIOR CLAY-WORKING MACHINERY.

Office and Works, 59—63 Center Street,

Cleveland, Ohio, Jan 23. 1901

Cleveland Jan 20. 1901.

To the Officers, and Members of the "American Ceramic Society.

Gentlemen:

Please accept my resignation as a member of Your Society, as I have entered upon another line of business from that in which I have been engaged in the past. Should I ever re-enter this line again I would consider it a great favor to be re-admitte as a member of Your honorable Society.

Yours very respectfuly

Gustav J. Holl.

Society membership was intended for people whose daily concern was ceramics and required rigorous standards to join. This early member wanted to keep his friendships in good repair in case he got back into the business.

discussion of the subjects after the formal presentation at the annual meeting. Ironically, just the opposite situation would be observed 30 years later, when Society officers complained that too many of the papers being read were last-minute rush jobs, and proposed that only those papers submitted two months in advance for editorial review and printed in the *Journal of the American Ceramic Society* could be read at the annual meeting.

Membership in the Society during its first decade was confined to those individuals who applied for admission and were deemed by the current members to meet their fairly rigid standards. As Orton later noted, "the number of men who could qualify educationally and on the basis of their technical achievements in ceramics were comparatively limited."

Even so, initial growth of the Society was remarkable. The second annual meeting, held in Detroit in February 1900, found that in just one year membership had grown from 22 charter members to 56 members (23 members and 33 associate members). Annual dues were $5 for members and $4 for associate members.

In the 19th century, ceramics was a profession usually learned on the job. In this photograph, the founders of the Society are shown with colleagues as very young men in what was titled "The First Class in Ceramics at Weller's Pottery, Zanesville, Ohio." From left, front row: Samuel Geijsbeek, J.W. Wolfley, Carl Giessen, E.F. Braddock, D.C. Thomas, E.J. Jones, Edward Orton, Jr., Elmer E. Gorton; in doorway, left: S.A. Weller; right, W.A. Long.

By the time of the seventh annual meeting, in 1905 in Birmingham, Alabama, the Society had 166 members, was experiencing an income of more than $2,000 annually and had gained prominence among scientific and engineering societies of the world. Visitors from England, Germany, France, Sweden and Japan were in attendance that year and were so impressed that they affiliated themselves with the Society.

Clearly, the Society's progress as the technical voice of the ceramic industry was little short of amazing. In just a few years *Transactions of the American Ceramic Society*, which published all the papers presented at its meetings, was recognized as the most important scientific ceramic journal in the United States and probably the world.

Recognizing that the Society was becoming an institution, the membership voted to proceed with the Society's incorporation. The articles of incorporation that created the not-for-profit American Ceramic Society were signed by Ohio's secretary of state on March 25, 1905.

That year at the annual meeting, the Society reported 33

EDWARD ORTON, JR. (1863-1932) RENAISSANCE MAN

No figure in Society history commands as much respect as Edward Orton, Jr., the organization's guiding spirit. He was not only a founder and for nearly 20 years the first general secretary of the organization, but also a giant in the field of ceramics. He was the leading force in every facet of the operations of the Society and is universally recognized as having had more influence on the growth and fortunes of the Society than anyone else.

He was born in 1863 in Chester, New York, the son of Dr. Edward Orton and Mary Jennings. The younger Orton studied mining engineering at The Ohio State University, graduating in 1884. His senior thesis was entitled, "Plans and Specifications for a Fire-Brick Factory." His interests were not all scientific, however. He also founded what is now The Ohio State Marching Band.

From 1884 to 1893 Orton worked as a chemist for a variety of iron and coal mines, blast furnaces and clay plants. In 1888 he married Mary Princess Anderson of Columbus, Ohio, who died in 1927.

It was while he was serving as the superintendent of an unprofitable brick plant in 1890 that Orton became convinced of the need to bring scientific principles to bear on his industry.

Orton was frustrated because he could find no technical literature about clay working — at least in English — that would assist him in solving production problems at his plant. At that time problems in ceramics were

U.S. members, one foreign member (William H. Zimmer of Coburg, Germany), 126 U.S. associates and 16 foreign associates. The Society's bank account had a balance of $615.62.

During the Society's early years two meetings were held each year. The annual meeting held in February was dominated by the formal presentation of papers and the management of Society business. The Excursion Meeting, a summer event, became, as one Society chronicler put it, an

> informal coming together of congenial professional spirits, with some attractive ceramic center as the objective, to combine shop visiting with innumerable meetings of numerous groups, where confidences can be safely exchanged beyond the reach of the reporters' ears.

After the creation of divisions beginning in 1919, additional meetings were held by the various divisions and sometimes by geographic region, or "Section." The last summer Excursion Meeting on record was in 1934 as a joint gathering with the Electrochemical Society. Fall division meetings began in 1930 with the Glass Division's first get-together at Cove Point, Maryland. Eventually the growing enthusiasm for meeting once a year in smaller groups with specific shared interests took precedence over the larger Society-wide summer tours.

Gradually the Winter Meeting was moved to spring and is currently in April or May. But even as the meetings' structures continued to evolve, the gathering of "congenial professional spirits" was prized as one of the most valuable aspects of Society membership.

For the first decade or so of the Society, it was standard practice for the annual meetings to alternate between business sessions and those in which papers were presented. The sequence usually was determined by the amount of pressing business and the discretion of the governing Council (later called the Board of Trustees). These activities were punctuated with visits to plants, studios, mines and quarries, and other business operations in the vicinity of the gathering.

THE GROWING YEARS

During the Society's 11th meeting, in February 1909 in Rochester, New York, long-simmering frustrations held by some members finally boiled over. For years complaints had been voiced in private that too few of the papers being presented and subsequently published were of practical value. Some members felt that many of the papers were too close to pure science and were well beyond the understanding of factory personnel.

solved by trial and error. Chemistry and physics were rarely employed and, in fact, were distrusted by many old-timers in the brick industry. This need for technical information led Orton to found the ceramic engineering program at The Ohio State University in 1894 and to take a founding interest in The American Ceramic Society.

He lobbied the state legislature on behalf of his so-called "Mud Pie Bill," and once money was appropriated for a department of ceramics and clayworking at OSU he became its first director, demanding that students engage in a rigorous two-year course of study that would expose them to all aspects of science as it applied to the ceramic industry.

Asked by some students why a potter should waste his time studying brickmaking, Orton fired back: "That is the main problem with the clayworkers of the present. They are too narrow and one sided. We all need to be forced to consider broader fields of action than our own, else we soon lose the power to advance even in our little specialty."

He served as state geologist of Ohio from 1899 to 1906, spending some time in Rocky Mountain National Park conducting a survey of Mills Moraine in the Long's Peak area. It was recalled by a summer resident of Estes Park who joined Orton's students in the project that they wanted to name a peak in honor of Orton. The professor declined, offering as a candidate his father, Edward Orton, Sr. Mount Orton first appeared on a map of the park in 1911 and was duly recorded by the U.S. Geographic Board in Washington.

About that same time Orton made this prediction:

> *A chemist in a pottery or glassworks is still as much an anomaly as he was at a blast furnace of a generation ago. But the brickworks and potteries of another generation will look back with wonder and marvel how it was possible to have struggled unaided so long, with the help at their doors for the asking.*

The first secretary of the Society, Orton served without pay for 20 years. When the United States entered World War I in 1917, Orton already was 53 years old — long past the point where he was required to serve his country in the military, or even to consider such service. He had his work with the Society, his teaching position and, in addition, operated the Orton Pyrometric Cone Factory in Columbus. He had every excuse for sitting out the war.

But by all accounts Orton was a patriot in the purest sense. He volunteered for service in the Army and was initially rejected because of his age. Eventually his persistence, enthusiasm for military service and his

In 1914, Edward Orton, Jr. requested to attend the first Plattsburgh Training Camp for civilians and was refused because he was more than 50 years old. However, his persistency, his enthusiasm for military service, and his excellent physique and keenness of intellect caused the War Department to change its decision and to permit him to attend and complete the course at the Plattsburgh Training Camp in the summer of 1915. In 1916, he took part in the drafting of the National Defense Act. On January 5, 1917, he was commissioned a Major in the Officers' Reserve Corps and was called to active service in the Motors Division of the Quartermaster's Corps on May 9, 1917. Major Orton saw service in the Quartermaster's Corps in Texas in 1917 and was later [in 1917] brought back to Washington at the request of General Chauncey B. Baker and put in command of the Engineering Division of the Motor Transport Corps. He was promoted to the rank of Colonel in 1918.

excellent physique and intellect won over the War Department. In the summer of 1915 he was assigned to a training camp in Plattsburgh, New York and subsequently was commissioned as a major in the Quartermaster Corps.

There his experience and intelligence quickly gained the respect of Army brass, who put him in charge of an effort to produce a truck to Army specifications. Orton called together the country's best automotive engineers and insisted that they ignore the usual competitive pressures and pool their resources to produce a standard motor vehicle.

"He had no time to listen to sales talk or to quibble over minute details," reports one of his biographers. "The questions with him seemed to be: Which is the best design, and what are the possibilities of the getting the desired number of cars or trucks within a given time? He dominated the designers, was irritated by unnecessary delays and much augmentation, and spent days and nights on the motorization of all departments of the United States Army."

The result was the Liberty motor and truck, used by the military in Europe and for several years after the end of hostilities. For his efforts Orton received the Army's Distinguished Service Medal. After the war he was made a brigadier general in the Army Reserve Corps, and was known as General Orton to Society members ever after.

He and others in science and industry drafted the National Defense Act of 1916, which was passed by Congress without change, establishing the National Research Council.

Even while he was in uniform, Orton was nominated for the presidency of the Society. The nominating committee felt that he had earned the post, which was largely symbolic. Orton didn't see it that way.

"Your remark that the office of president does not involve any great responsibility or expenditure of time or labor probably applies to the Society as it would exist with anybody else as president but myself," Orton wrote in declining the nomination. "I seem to be unlucky in the fact that I have never yet had a job in any organization that did not take time and labor and plenty of it, and I know myself well enough to know that if I were president of A.C.S. I would be spending a lot of time and energy on it."

He received an honorary Doctor of Science degree from Rutgers University in 1921, and the degree of Doctor of Laws from Alfred University in 1931, the same year The Ohio State University conferred on him the professional degree of Ceramic Engineer.

He served as president of The American Ceramic Society during 1930–1931. Those who knew Orton described a man of perseverance and potent personality, a demanding taskmaster who nonetheless realized his goal to create a "delightful

That year the complaints were made aloud from the floor by Society members whose interests were largely commercial and who had little faith in science and even less desire to accept the research interests of many ceramic scientists.

In rebuttal, the scientists and engineers emphatically insisted that the new knowledge contained in the papers was indeed of practical value and that the die-hard clay workers should acquaint themselves with it.

This debate has become the longest-running battle within the ranks of the Society, and is still a source of vigorous discussions a century later.

Meanwhile, the Society continued to grow. After years of contention and many misgivings on the part of scientific purists among the membership, the Society in 1910 established a contributing member class for those companies that wished to participate in the Society's efforts. After 1918 these were referred to as corporate members; today they number 200. The first company to join the Society as a contributing member was Pittsburgh Plate Glass Co., now known as PPG Industries, Inc., which joined in 1912.

Student branches were established in 1915, after the Society was petitioned by the ceramic department at The Ohio State University. Shortly thereafter student branches were formed at what was then the New York State School of Clay-Working and Ceramics at Alfred University, and at the University of Illinois, Rutgers College and Iowa State College.

Also in 1915, it was proposed that local sections be established in different parts of the country. Several years before, a group in Beaver Falls, Pennsylvania, had organized the Beaver Ceramic Society, and this became the first local section. It is now known as the Pittsburgh Section.

By 1918 sections were established in northern Ohio, central Ohio, New England, New York state, Chicago and St. Louis. Central Ohio established a section and the New Jersey Clay Workers' Association became the Eastern Section in 1919.

The value of the sections was immediately felt. This was before the era of commercial airlines, so traveling to a national convention could mean three days on trains for members on the West Coast. The sections allowed for more frequent meetings closer to where members lived and worked.

Curiously, the advent of air travel and rapid transit has not diminished the importance of the sections in many areas. Society officials today confirm that for many members, sectional gatherings are their only opportunity

atmosphere of good fellowship existing in but few technical societies." He insisted on continuous hard work on the part of the membership and, step by step, led the organization to the position it now occupies in the industrial world.

Orton died in 1932; his will established the Edward Orton, Jr. Ceramic Foundation to advance the ceramic arts and industries. The Foundation, based in Westerville, Ohio, still exists today. In eulogizing Orton, his close friend, the Rev. M.H. Hichliter of the First Congregational Church in Columbus, said:

> *A rebirth of intellectual honesty, a fearless scientific search for facts, a willingness to speak the truth as God gives a man to see the truth, an unselfish devotion to his family and to the public good, a deep instinctive sympathy with the oppressed, the neglected and the needy classes — these are the marks of this happy warrior who being dead yet speaketh.*

ORTON DOES IT ALL

In the Society's first two decades, the remarkable Edward Orton, Jr. served not only as secretary but ended up doing just about anything else that came his way in serving the organization.

At the 14th annual meeting held in March 1912 in Chicago, Ellis Lovejoy, then treasurer, made these comments when called upon to make his report just after Orton had delivered his own exhaustive analysis of the Society's status:

> *Mr. President, the report of the secretary is so complete that the report of the treasurer is little more than perfunctory. As a matter of fact, the secretary has taken most of the work of the treasurer on his own shoulders, leaving the treasurer the responsibility and the honor. The honor is great, but the responsibility is very small. As a matter of fact, as fast as the money comes in they spend it, and sometimes they spend it before it comes in, so there is practically no responsibility. I am glad to have been treasurer this year. I have learned how to do it.*

(Right) In 1936, at the Society's first rented office space on North High Street in Columbus, a professorial portrait of Gen. Orton (left) dominated a wall studded with photographs of Society founders and other scientists.

Mrs. Edward Orton, Jr. (Althea, his second wife) was honored at a student reception at Rutgers University, which had conferred an honorary Doctor of Science degree on Orton, who was known as General Orton for his postwar rank of brigadier general in the Army Reserve Corps.

ORTON STEPS DOWN

While camaraderie and professional respect had been core values of the Society from its beginning, there were too many proudly held views and strong characters for member relations to be completely amicable. It was during the war years and Orton's absence that internal friction within the Society reached a flash point. Not unexpectedly, the main player in this incident was Ross C. Purdy, whose devotion to ceramics often carried him to excess. Both Purdy's devotion to the Society and his combative personality would figure prominently in Society business over the next 25 years.

The impetus for one particular outbreak of Purdy's wrath was his dissatisfaction with the performance of Arthur S. Watts, a past Society president and, like Orton, a faculty member at The Ohio State University (and, as luck would have it, Purdy's future brother-in-law).

With Orton in the Army, Watts had agreed to take over the duties of secretary and editor. When it was proposed that Orton once again be elected to the secretary's position — even though in his absence things still would be run by his hand-picked choice, who was Watts — Purdy exploded.

Purdy, then chair of the membership committee, fired off a scathing letter to Orton:

I submit to you in my usual frankness that I will oppose your election if it means that in your absence Mr. Watts will serve as secretary pro tem. You well know the loyalty with which the members stand by you. Rather than cross your wishes they would submit to much to which otherwise they would offer stiff objections.

Purdy, though, had no qualms about airing his "stiff objections." In his extraordinary letter to Orton he wrote that:

I am able to keep separate my personal hatred of you and my professional respect for you. . . . When I deem it against the welfare of the Society to see to hear technical presentations, providing a forum for the exchange of ideas, for developing professional contacts and for enhancing professional growth.

In the Society's early years, many regarded it as "the scientific division of the N.B.M.A.," a designation that did not suit Society members. For all intents and purposes, the Society was an independent body that met at the same time as the Brick Manufacturers simply for the convenience of individuals who were members of both organizations.

In 1911 the Society for the first time met independently of the N.B.M.A. By 1916 the break was official — it was determined that it was no longer necessary for the Society to hold its meetings at the same time and place as the older organization. Though many individuals remained members of both groups, it was thought that The American Ceramic Society had grown to the point that it could strike out on its own. It is an interesting historic note that the roles were reversed in 1931 when the N.B.M.A. merged with the Heavy Clay Products Division of The American Ceramic Society to form the Structural Clay Products Division.

The growth of the Society, while steady, was not without occasional setbacks in these early years. At the 1915 meeting, for example, it was reported that 31 memberships had been lost in the previous 12 months. While membership continued to grow, then, as in the years since, the loss of any members was a source of concern. Orton, in his secretary's report, conjectured that, because the Society's activities were limited to one annual meeting, a sparsely attended summer meeting and the issue of one volume of *Transactions* every fall, "under those conditions it is a hard proposition to keep up the keenest interest."

THE GREAT WAR

As the Society reached its 20th year, it was an obvious success. It had grown from membership of 22 to nearly 1,000 and represented industries having an aggregate output valued at $450 million annually in the United States alone.

Even more remarkable, membership had doubled in the previous two years, in part because of the emphasis on scientific development during World War I. The Great War brought an end to a world ruled by traditional means and materials in many fields and a beginning to the leadership of science in almost every aspect of human endeavor. Ceramic industries and research flourished.

The war found Society members active in a variety of efforts to support the military and to ensure a

> *not wavering one whit in my regard for your past services or in my professional loyalty to you. I consider it my duty that you withdraw your nomination.*

Getting to the bottom of this incident after 80 years is difficult, in part because even then Society members could not agree on just where the truth was to be found. Clearly, some members were dissatisfied with Watts' performance, but others observed that, in fairness, it should be admitted that anyone was going to look incompetent after 20 years of Orton's brisk efficiency, and that things were not nearly as bad as Purdy suggested.

Watts countered that Purdy's attack was a personal vendetta against both himself and Orton for imagined slights; in retrospect it is quite possible that Purdy was cannily laying the groundwork for his own ascension to the position of the Society's first paid secretary a few years later.

Meanwhile, Purdy was lobbying the Board to hire a full-time secretary and to make other sweeping changes in the Society — publication of a nontechnical periodical, greater cooperative activities with other national organizations, more active industrial work, changes in full membership requirements that would open up the Society, more local sections and the creation of industrial divisions. It was feared that these suggestions would move the Society away from its status as a purely scientific organization, and they were vigorously opposed by the organization's old guard.

Orton, still in uniform, decided it was in the best interest of all to request that his nomination for secretary be withdrawn:

> *So far as I can learn, the number of those who hold the view that my nomination for secretary should be recalled is rather small. Nevertheless, the fact that there are any at all is sufficient to cause me to consider the matter very seriously. . . . It seems to me better to at once end this ambiguous situation by withdrawing my nomination for secretary for another year, and at the same time to resign from even that nominal connection with the secretaryship which I have held since I entered the Army. This I hereby do.*

Despite howls of protests from many members, Orton held firm. After nearly 20 years, he was stepping down.

Orton then wrote to Purdy:

> *Every time you write me you cite a new list of slights and affronts which I have put upon you, and each time I am amazed at what you have treasured up and*

what construction you have put upon it. I probably have been blunt and self-centered, and unsocial and oblivious of some obligations that to you seem rather fundamental to friendship. But that's all! I could not be happy myself if I were feeling as you think I do. It is foreign to my nature and disposition to quarrel or to hate. So, insofar as you attribute unkindness to me, forget it! Nothing doing!

With Orton out of the running, Charles F. Binns of Alfred University was elected secretary of the Society. Still, Orton's many supporters were livid, and Purdy felt it prudent to write Orton, requesting that the founding father explain to the membership why this was all for the best.

Whatever Ross Purdy's motives and methods in this incident, in retrospect it is clear that he possessed an all-encompassing vision for The American Ceramic Society and was determined to see it implemented. Many members actually liked Purdy's style.

"I am certainly delighted to find that the Ceramic Society is taking a new lease on life," member Warren Emley wrote to Purdy.

I must confess that for the past few years I have been paying my dues through a sense of loyalty rather than because I felt that any benefit was to be derived from the Society. I have, however, felt very loath to appeal to the lime industry because I could not see how the activities of the Society could be of any particular benefit to them. . . The thought has occurred to me that there is a need of a clearing house of technical information among lime men and I am wondering if the Ceramic Society could be used in this capacity . . .

Meanwhile, under Purdy's aggressive leadership of the Membership Committee, nearly 200 new members signed on in 1917. ▲

In 1930, a dapper Orton, along with other Society members and their wives, visited Stoke-on-Trent, England, the birthplace of the great ceramist Josiah Wedgwood on the celebration of the 200th anniversary of his birth, for a series of social and professional activities with ceramists from all over the world.

CITY OF STOKE-ON-TRENT.

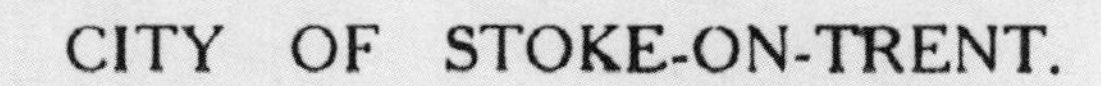

The Lord Mayor (Alderman George H. Barber)

requests the pleasure of the

company of

Gen. Ed. Orton & Mrs Orton

To DINNER at the Town Hall, Stoke-on-Trent,

On Thursday, the 22nd May, 1930, at 7-30 p.m.,

on the occasion of the

Visit of the Foreign Representatives of the Ceramic Society.

steady flow of ceramic materials and products to the country's factories.

Orton recognized the potential for ceramists. He organized the Society's committee on military and economic preparedness, addressing first the war effort and, secondarily, economic matters affected by the war, including needs and opportunities in the post-war period.

Then, in 1917 at the age of 53, Orton enlisted in the Army, which gave him the assignment of creating a new truck for the transportation of men and supplies. Creating a brain trust of the finest automotive engineers in the nation, Orton spearheaded the development of the Liberty motor and truck. His efforts won him the Army's Distinguished Service Medal. Later he was made a brigadier general in the Army Reserve Corps and thereafter was referred to as "General Orton" in Society correspondence.

Meanwhile, other Society members were engaged in a variety of war-related efforts through various committees. Ross C. Purdy headed a subcommittee on abrasives and organized the makers of abrasive wheels, which were essential for the fine grinding of optical glass for range finders, gun sights and other instruments needed for the war effort.

Even more pressing was the development of new spark plugs for airplanes for the world's first aerial combat and truck engines for the first motorized war. Both kinds of machines, still in their earliest development, were subjected to extremes of heat and cold. The committee on electrical porcelain, under the leadership of L. E. Barttinger, completed an elaborate study of the problem, which resulted in the development of a new porcelain composition that met the rigorous requirements. This information was passed on to spark plug manufacturers. In addition, a new insulation for magnetos was created.

The hostilities meant that high-quality German porcelain was no longer imported to

A plate from the Red Book, a lavishly illustrated customer catalogue used by Pittsburgh Plate Glass at the turn of the century, showing a great pane of glass as it was made at that time, bears this inscription: "The roller passes over the red and paste-like glass; a workman, with his eyes fixed on the fiery substance, skims off all apparent defects with a bold and skilled hand; then the roller falls and passes off. . ."

CAN'T GET NO RESPECT

Do ceramic schools receive the assistance and encouragement from the manufacturers of clay products they deserve?

I fear we must admit they do not; look around you, think over the clay factories known to you, and how many of them are employing one of these technical students? What encouragement is there for a young man to take up the study of clays if employment can not be found in his chosen profession? To quote from a letter received from the director of a large technical school, in which he says, 'The average young man does not realize the possibilities in the clay-working industries.'

We might ask, does the average manufacturer realize the great possibilities? If he did, the man who has made technology his special study would be placed in the factory to assist the practical man in working out the problems which are causing constant losses.

— *Francis W. Walker, in his retirement address as Society president, 1905*

ROSS C. PURDY (1875 - 1949) SOCIETY CHAMPION

Perhaps the most colorful character ever associated with the Society, Ross C. Purdy had a zest for his work that could put a twinkle in his eye — or send him into a tirade if he believed that American ceramics or the Society itself was slighted in any way. He had a reputation for charming every lady he met. He was married nearly 50 years to Myra Jane "Mother" Purdy (far right in photo below).

Ross Purdy was The American Ceramic Society for 24 years, from 1921 to 1946, and was, by all accounts, as colorful a character as the group has ever produced.

Purdy was born March 3, 1875, one of five sons of Andrew and Mary Elizabeth Purdy. A native of Jasper, New York, Purdy attended school in Pennsylvania and New York. In 1895 he began his academic career at Syracuse University, transferring after one year to The Ohio State University.

After three years at the university, Purdy left to work for the Mosaic Tile Co. in Zanesville, Ohio. In 1902 he and his new wife, Myra, moved to Macom, Illinois, where Purdy developed a line of stoneware specialties. He eventually left this job to take a part-time position that would permit him to return to school for a ceramic engineering degree.

In 1906, after a brief instructorship at the University of Illinois, Purdy returned to The Ohio State University, where he served as Edward Orton, Jr.'s assistant in the College of Engineering. Purdy, who had always been active in societies and ceramic organizations, soon became an early member of The American Ceramic Society.

Energetic, combative and, in various circumstances, as charming as he could be argumentative, Purdy served as Society president in 1909. He was the first editor of Transactions of the American Ceramic Society *in 1911, the first published proceedings from the Society's annual meetings.*

In 1921 the Board hired Purdy as the Society's first full-time employee. He took over the position of general secretary, held for nearly 20 years by Orton, whom Purdy appeared to regard as a rival.

When Purdy came aboard as general secretary, membership was about 450. By the time he retired it was nearly 3,000. If Orton steered the Society through its infancy and childhood, Purdy brought it from adolescence to adulthood.

As the long-time editor of the Journal of the American Ceramic Society *(at that time the general secretary also was responsible for the Society's publishing ventures), he was known for his insightful editorials. Reviewing Purdy's work in 1976, an admirer noted, "They were timely when written and many could be republished today without any loss in significance."*

In print Purdy's editorial voice ran the gamut, sometimes chiding, sometimes cheering the membership. Whatever tone he took, Purdy's objectives were always in the Society's best interest.

When space allowed, Purdy, the proud grandfather, was known to include photographs of his grandchildren in the Bulletin.

Purdy was an early champion of the creation of sections and divisions, maintaining that "our programs are getting too long now and the subject matter is too diversified to make it

the United States. Charles F. Binns, who had come to the United States from Royal Worcester Porcelain Works, where he had spent 26 years making porcelain for the British crown, headed a committee charged with improving the production and quality of American porcelain. The committee hoped not only to keep those materials available during the war but to compete with the German industry after the armistice.

DIVIDED WE STAND

During its early years the Society at its annual meetings convened as a whole — a single daily session was attended by all members and papers from all portions of the field were read and discussed. But the very success of this approach assured that it could not last. There was too much to say and too many to listen for single sessions to continue.

As the Society grew, meetings were divided into two sessions — one devoted to the heavy clay products industry, the other to whitewares. This division gave each member an opportunity to choose the session that more closely mirrored his own professional interests. But even as early as 1910 some members were calling for separate sessions for each of the different segments of the industry.

By 1918 the Society had largely outgrown the notion that every member possessed a broad interest in all kinds of ceramic activity. Presenters didn't want an audience of strangers to their topics. Specialization was called for. As the *Journal of the American Ceramic Society* noted nearly 20 years later, "even the technical men from our schools and laboratories were loath to read papers on glass and enamels to audiences made up practically of persons not interested in the subject being presented."

The 21st annual meeting, held in February 1919 in Pittsburgh's Fort Pitt Hotel, saw the creation of professional divisions within the Society, the first major change in the Society's government since its inception. That meeting marked the first time separate division gatherings were held — for enamel, glass, terra cotta and refractories. Meanwhile other divisions — devoted to abrasives, brick and tile, floor and wall tile and whitewares — were being organized.

With the advancement of science and industry over the years, new divisions were created and others metamorphosed. The Heavy Clay Products and Whitewares divisions began functioning in 1921; they were quickly followed the next year by the Art Division (which later became the Design Division).

None of this occurred without considerable controversy. "The fear and expectation was quite general that

worthwhile for one's attention to be here all the time." His enthusiasm also encouraged the Federal Bureau's interest in ceramics.

Purdy was recognized for his tireless work to promote the Society when he was presented for Honorary Membership during the 44th Annual Meeting in 1942. It was a difficult time for the Society; many members had dropped out or entered the military, and the Society was spending more money than it was bringing in. Purdy did what he could to help, paying his assistant out of his own pocket.

"He kept the Society alive," observed past president Jack Nordyke. "It could have died right there. He was most respected for that."

Purdy's generosity touched the Society in other ways as well. Employees of the Society recalled years later with great fondness that it wasn't unusual for Purdy to stop at a bakery on his way to work and lay out several pies for the office staff. His greatest gifts came upon his retirement, when he contributed many of the books, photographs and invaluable ceramic pieces that are now part of the Ross C. Purdy Museum of Ceramics collection.

Purdy's loyalty to the Society extended to the American ceramics market, of which he was a staunch defender. The general secretary refused to eat off foreign-made plates, and when he discovered during one Society annual meeting that the banquet hall tables were set with Japanese dinnerware he delivered an ultimatum: If there wasn't American ware on the table in an hour, there would be no banquet. Panicked hotel employees raided every other hotel in town to get enough American ware for the event. It didn't all match, but it was American.

Purdy received dozens of honors during his lifetime, including the first Albert Victor Bleininger Award. Presented to him by the Pittsburgh Section of The American Ceramic Society, the award recognized Purdy for Distinguished Achievement

(Above) Inscribed "Presented to Colonel Ross Coffin Purdy by the American Ceramic Society" the crystal saber with a ruby-glass handle was originally presented to Purdy during a western frontier gala at the Society's 1935 Annual Meeting. Today the saber is part of the collection in the Ross C. Purdy Museum of Ceramics.

in the Field of Ceramics. Past Society President E.H. Fritz commended Purdy, who Fritz said "was more than any other individual responsible for the outstanding growth of the ceramic industry."

More than just an advocate for the Society, Purdy was an ambassador who earned the respect of colleagues around the world. Upon learning of Purdy's passing in 1949, Rudolph Barta, honorable general secretary of the Czechoslovak Ceramic Society, remembered fondly:

> . . . when he firstly [sic] came here, we expected to see a pedantic scholar. But what was our surprise when a gay and witty companion appeared, bounding everyone with his personal charm.
>
> . . . Here in our country we shall think of Dr. Purdy not only as a famous scholar, man of merit, well-known organizer of American Ceramists and our friend, but, first of all we shall remember him as a kind man.

(Below) The office of the general secretary in the 2525 North High Street building was filled with paintings and objects celebrating the history of ceramics, especially American ceramics.

PURDY — EDITORIAL WRITER

The hallmark of Ross Purdy's administration was the belief that ceramics had an unlimited future, but only if ceramists devoted themselves to the promotion of their industry with the same enthusiasm they brought to bear on scientific problems.

There was resistance to this notion, which Purdy countered in his writing. The following are examples of Ross Purdy, editorial writer, at his best:

> Ceramic wares are among the oldest of man's handiwork and they were created because of man's urgent need for them. Because they have been necessities these many centuries no amount of effort has been required to sell all that a ceramic manufacturer produced. There were no, or at the most very few, competitive products. There were no 'heavy duty' specifications to meet.
>
> For so many decades have our ceramic ware producers eased along that the present-day keen competition from other products and the absolute need for qualities to withstand heavier stresses, higher temperatures, higher voltages, higher efficiency and more severe usage have caught us unaware. We do not know what we should do or how to do it. The only thing we are sure of is that business is slipping away from us and that producers of competitive products are going stronger with each succeeding year, month, week and day.

the federation would not survive longer than it would take the divisions to become sufficiently strong to go off by themselves," observed one Society historian.

An early report by Orton on the possibility of creating divisions underlined

> the understanding . . . that the Society is not to be divided physically into sections meeting in different places, but that the division is a sort of intangible one and that the section leaders are, in fact, assistant secretaries, each section leader desiring and trying to keep in contact with the men in the Society that he knows are interested in that field and trying to stir up debates, papers and program material — in other words, to build off the work of the Society, each in his own field.

In the end Orton was proven right. The divisions became the means of the Society's gaining strength — though not without creating a few headaches for administration.

WHOSE SOLE BUSINESS IN LIFE

By 1921 it was obvious to everyone that the business of the Society was now so complex and overwhelming a task that it was no longer feasible to place all this responsibility on unpaid volunteers.

"The employment of a full-time secretary-editor has been contemplated for several years," the retiring president, Homer F. Staley, told members attending the 1919 convention in Washington, D.C.

> We should have one man whose sole business in life shall be to attend to the affairs of The American Ceramic Society. The secretary will be kept very busy indeed maintaining a voluminous correspondence, in overseeing the business affairs of the Society, in editing the *Journal*, in writing editorial and in securing advertisements.
>
> The Society has few men who can fill the position properly, and probably most of these will not care to consider a position so insecure of tenure and exposing the occupant to so much criticism. The best man available should be secured; and then the members of the Society should remember that while he will be the servant of the majority, he will in no sense be subject to the criticism or wishes of individual members.

Thus the office of secretary was created by the Board of Trustees. To remove the position from petty politics and to provide the holder of the job with at least the prospect of continuous employment, the

What are we going to do about it?

The present answer is that we will fight among and with ourselves, lessening our chances, while our competitors grow stronger. Face brick, common brick and structural tile will openly fight each other. Chinaware producers will stare at the red ink in their books and growl at the glass man. The concern with larger resources will dictate lower and lower prices in the vain hope of killing off his weaker brothers in trade. We will not cooperate throughout our respective trade associations and our several trade associations will not cooperate with each other.

It is high time that ceramic manufacturers, one and all, federate together in a Ceramic Research Council to study the situation, produce information and make applications. . .

We have eased along altogether too long. We must get together.

. . . One notable result of the European War has been the marked change in the status of industrial research in this country. Before the war research was looked upon as an interesting side issue of academic life, very indirectly related to factory management, but from the financial standpoint to be considered a non-essential luxury. During the war, the value and necessity of research were well proven, and consequently it has received a tremendous amount of advertising. Today industrial research is not only respected, it has reached the enviable but precarious position of being a fashionable investment.

The money is available; the question is how shall it be invested wisely?

. . . We should not rest our case with the middleman. The middleman is not singlemindedly for American ceramic ware or indeed for ceramic ware at all. It matters not to him what he sells. He meets demands. The manufacturer must create those demands with the ultimate purchaser.

Many members believe that the Society has done and is doing very well. Compared with past performance this Society has shown a steady progress. Compared with what it should be in power to do and in present activities, this Society is so far short as to warrant calling it a 'piker.'

Your secretary is not pessimistic or discouraged. He is not a grouch. His ambitions for the Society have not been realized. He wishes that recognition of this shortage shall be shared by his fellow members. . . .

This Society is not keeping pace with its opportunities. It must do more than merely initiate; it must lead in cooperative enterprises if the members are to get the greatest

possible returns from their invested efforts. The Society is not developing nearly as rapidly as are the needs for the sort of service for which it was organized to render. ▲

(Left) Purdy made a call — perhaps to check up on one more piece of Society business — before a testimonial dinner in his honor in Buffalo, New York, in 1946, the year he retired.

(Below, left) Many of the ceramic pieces collected by Purdy and his wife were given to the Society after his death and today make up the heart of the Ross C. Purdy Museum collection. Known to Society members as "Mother Purdy," the general secretary's wife welcomed Society members and staff to the Purdy home for countless meetings.

(Below, right) In 1942, Purdy's bust was cast in glass to commemorate his honorary membership in the Society. Created by Frederick Carder of the Steuben Division of Corning Glass Works, the seven-inch, one-piece bust is displayed in the Ross C. Purdy Museum of Ceramics.

secretary was to be chosen by the Board rather than by a popular vote of the members.

The secretary's duties as described at the time were to assist the Board of Trustees, the committees, local sections, divisions and other segments of the Society, and to hold annual meetings, publish the *Journal of the American Ceramic Society*, cooperate with other societies and organizations and stimulate research through cooperation with the National Research Council.

The man for the job was obviously the aggressive and dedicated Ross C. Purdy, who was frightened neither by insecurity of tenure nor of criticism. Purdy, The Ohio State University ceramics professor, was appointed to this new position on July 11, 1921, with a salary of $4,000 per year and a yearly allowance of $2,000 for clerical and travel expenses.

At the same time, The Ohio State University offered free of charge an office in Lord Hall to serve as the Society's headquarters. The Board accepted the offer and Purdy immediately set up offices there.

Simultaneously with the hiring of a professional secretary, the Society made a few changes in its governing policies. Before 1921 trustees were elected by the membership at-large. Thereafter, each division elected one trustee to serve a three-year term as its representative on the Board. This change was adopted to permit each recognized branch of ceramic technology equal representation in the governing of the Society.

If Edward Orton, Jr. dominated the Society during its first two decades, Purdy became its unmistakable leader for the next two decades. Strong-willed and visionary, Purdy believed that for the ceramic industry to take its rightful place in American society, it would have to embrace both the purely scientific and, at least to some degree, the promotional.

Purdy urged Society participation in ceramic exhibitions, a move dismissed by some members as crass marketing efforts. He urged that the Society assume a more assertive role in encouraging the development and marketing of new ceramic products. From his position as editor of the *Journal*, he endlessly cajoled, berated and challenged the members to see things his way.

And when they didn't — as when in the early 1930s the Board of Trustees informed him that it would not authorize the expenditures for the Society to participate in a ceramics exposition — Purdy threatened to defy the board's authority and proceed anyway.

Undoubtedly many who served on the Board of Trustees during the Purdy era found their secretary to be frustrating, even infuriating. Significantly, though, he

SCIENCE AND PRACTICALITY

For much of its history, the story of The American Ceramic Society has been of the give and take — and occasionally outright antagonism — between what have been called the "theoretical" and "pragmatic" sides of the industry.

Initially, many clay workers were dubious about the value of scientific analysis and its application to their work. They preferred to rely on time-tested methods and their own instincts.

Scientists, such as Albert V. Bleininger, found that thinking short-sighted and self-defeating, and took to the pages of industry publications to argue their case. In a 1900 copy of The Clay-Worker, *Bleininger argued that scientific theory was, in the end, completely pragmatic:*

> *At the close of the nineteenth century, would it not be well to survey the status of our knowledge in regard to facts of the clay industry? To what extent is science ready to help us in our work at this date, or is all the talk of science mere sound of words?*
>
> *First of all, what is science? It is simply organized and sympathetic knowledge, the logical result of experience and work. All of us are scientists if we trace in what we do the current of the natural laws and to make our work conform to those laws.*
>
> *There still exists an impression that the scientist is a man whose world is a world of books, and of whose general knowledge is of the most impractical kind. Never was there a more grievous mistake. Modern science is entirely practical, and the scientist is a practical man wherever he may be at work — in the laboratory, the field or the shop. If he is not practical, if he is not willing to work hard, if he is not willing to make sacrifices, if he is not willing to be unselfish, he has no claim to the name of scientist.*

Twelve years later, then-president Arthur S. Watts applauded the technical accomplishment of the papers being presented by the Society, but cautioned against taking a strictly theoretical approach to ceramics:

> *Worthy of mention in connection with our more scientific articles is the desirability of drawing at least some practical conclusion from the data presented. . . . Few articles have appeared in our* Transactions *which have not a practical application, but many of them are not expressed in terms which the average clay worker will understand. If the practical application is woven into the article, he will see the*

importance of acquainting himself with data presented, which he might otherwise condemn as purely theoretical.

Ross Purdy took up the subject in a 1920 editorial in the Bulletin:

For conducting research two general methods are available, the empiric or engineering, and the theoretical or scientific. The empiric method . . . [starts] without any theory and by empiric trials collecting data for use in deriving an engineering formula or the basis of a new process of manufacture.

Empiric researches usually keep close to established engineering practice and are rather more certain to yield usable results than theoretical researches. On other hand, these results are not as spectacular and revolutionary in their effects as the results of the occasional successful research based on theory. As investments, empiric researches may be considered as safe; but, like all safe investments, they yield a comparatively moderate return on the money invested. Researches based on theory must be considered more hazardous, but they yield large returns if successful. ▲

Albert V. Bleininger (1873-1946), a charter member of the Society, was a German immigrant as a boy and graduated from The Ohio State University soon after Edward Orton, Jr. began the ceramics program there (Bleininger named his son "Orton Bleininger"). Later, he earned a Doctor of Science degree from Alfred University. His career in ceramics was long and distinguished, ranging from laboratories in the U.S. Geological Survey to the Homer Laughlin China Company. Probably the most recognized of his extensive publications is his translation of Hermann Seger's works, the first publication of the Society.

held on to his position. Even those who disliked him personally had to admit that Purdy worked tirelessly, ran a highly effective office and had a grasp of the big picture missed by other scientists and technicians who were wrapped up in their own specialized corner of the ceramics world.

Purdy was determined to expand the Society's reach, and he made two trips to Europe to establish The American Ceramic Society firmly in international affairs of the ceramic industry. On such a tour in the summer of 1928, Purdy led a delegation of Society members to London, Delft, Amsterdam, Berlin Dresden, Prague, Nuremberg, Munich, Lucerne, Paris and Limoges. The trip, he wrote in the *Journal*, was

> organized to meet an insistent demand by American industrialists and artists for first-hand information on ceramic markets, manufacturing processes, merchandising methods, general business conditions and such similar facts as are known to lead to increased business. The tour gives each member an opportunity to glean valuable information of actual monetary value to him and to his firm and at the same time enable his family to enjoy the social side of European travel with proper care and guidance.

The cost per person for this six-week adventure: $900. Purdy was not mistaken. It was so successful that it gained an almost legendary status in Society memory.

Though the Society continued to grow in prestige during the 1920s, it was also a time of financial instability. "We have been bothered considerably in recent years with our expenditures exceeding our income and we foresaw the financial troubles ahead, if this policy were not changed," R.L. Clare told the members in his presidential address of 1927.

Clare and the Board called for paring down all unnecessary expenditures and for cost-cutting efforts at the *Journal*, because publishing was taking the lion's share of the Society's budget.

Clare also called for a membership drive, noting that a regular annual increase should be expected:

> If such is not forthcoming, something is wrong, either in the quality of the service we are rendering, or in the way the affairs of the Society are being handled. I am decidedly against a strenuous membership campaign method that forces members on our rolls who have no real interest in our Society and who soon drop out. I believe that we should make our Society so valuable that it attracts new members.

It was a period of soul searching as members tried to come to grips with what the Society could and couldn't do to advance ceramics. The controversy crystallized in Purdy's determination to have the Society participate — even instigate — industrial and educational expositions that would present the industry and its products before the general public. Then, as now, the man or woman in the street had no idea just what was meant by the word "ceramic."

The Chicago Exposition of 1929 provided the opportunity for the first national general exposition of ceramic products held in America. From the pages of the *Journal*, Purdy enumerated its value to ceramics as an industry and a science. He was listing goals that have continued to be important to the Society even today.

- To bring various brothers of the big family of ceramic industries together.
- To develop a ceramic consciousness among the public.
- To demonstrate the importance and superiority of ceramic products in everyday life.
- To impress the public with the fact that America does not lag in quality, beauty and utility of ceramic products.
- To have something that — through the daily press and radio — a publicity campaign could be built around that would enable the ceramic industries to get a valuable message across to the public at a very low cost and in a very effective way.

"No trade association has nor probably ever will attempt such an exposition as will be held in Chicago," Purdy crowed. "It must be under the auspices of an organization interested in all ceramic lines."

In 1929, a group of English ceramists sailed on the R.M.S. Laconia to America to return the Society's 1928 trip to Europe, an excursion so luxurious and professionally productive that it became a fabled addition to Society lore. In France, the group posed on a sightseeing trip. The journey was the first of several that built good working relationships between European and American ceramists.

THE FEMALE PERSUASION

At the Society's first meeting, Maria Longworth Nichols Storer (Mrs. Bellamy Storer), founder of the fabled Rookwood Pottery, was made an honorary member.

For much of the history of The American Ceramic Society, women have been outsiders, a fact that reflects not so much the will of the membership as the condition of American society as a whole.

Women often attended the Society's meetings — as the guests of their husbands.

Even as late as 1959 at the 61st Annual Meeting in Chicago, the presence of a woman in a position of some visibility was an occasion of note.

"At least one entry in the annual student speaking contest is of the female persuasion," reported the Bulletin *that year. "There will be a girl [Frances Gaides of the University of California] among the student pages for the first time in the memory of modern members, and for the biggest break in tradition, the banquet speaker will be a woman" — advice columnist Ann Landers. "This is a contingency not covered in the Society's constitution, by-laws or rules."*

The following year the Bulletin *ran a photo of Leone P. Murphy, described as "the first woman in many years to receive a degree in ceramic engineering at the University of Illinois."*

Then, in a classic case of one-step-forward, two-steps-back, in that same issue job advertisements were grouped under the heading, "Men Wanted."

It wasn't until 1978 that a Board member noticed that the language of the installation ceremony read, "Gentlemen, you are about to be inducted as officers of The American Ceramic Society," and proposed revised language that would include both genders, and in 1985 "chair" replaced "chairman" to refer to all presidents and officers.

Nonetheless, over the years women have made their presence felt in too many arenas to name them all. Here are five that represent different roles: Maria Longworth Nichols Storer, a potter and business owner; Josephine Gitter, a manufacturer's representative and tireless chronicler of Society events; Emily Van Schoick, Society staff; Bonnie Dunbar, ceramist, astronaut and public celebrity; and Carol Jantzen, scientist and Society leader.

At the very first meeting of The American Ceramic Society, members voted to make Maria Longworth Nichols Storer (1849-1932) the first honorary member.

In 1873 she was a neighbor in Cincinnati of 12-year-old Karl Langenbeck (later a charter member and second president of the Society) when the boy received a box of china-painting colors from an uncle in Germany. Mrs. Nichols was fascinated by the process of china decoration, and by 1880 had opened the Rookwood Pottery where she controlled the entire production, from clay mixtures, colors, glazes and especially the firing temperatures of the kilns. Rookwood soon became world famous for its beautifully decorated vases, gardenware

Bonnie Dunbar earned degrees in ceramic engineering and a Ph.D. in biomedical engineering before becoming a NASA astronaut recognized internationally for her space engineering and science work.

That organization, according to Purdy, was The American Ceramic Society. But his plans fell flat.

Unfortunately for Purdy's plans, the Board of Trustees had doubts on two important points. First, the money wasn't there. Launching a series of ceramic expositions would cost money, and there was no guarantee of a quick return on the investment. Things already were financially tight, and there was no available seed money to get the project off the ground.

Second, there was serious philosophical opposition to having the Society go beyond its mandate as a scientific organization. Promotion wasn't part of the deal and might even violate the Society's tax-exempt status, some argued.

The arguments have never lost their currency as the Society has struggled with its role in making the public aware of ceramics and ceramic applications, and even of marketing the potential of the ceramic industry.

No doubt reflecting Purdy's irresistible influence, in 1929 then-President MacDonald C. Booze wrote in the *Journal* of the Society's potential to become the voice of the American ceramic industry:

> The interest of The American Ceramic Society in this matter is entirely in promoting the welfare of the industry at large. It has been necessary for the Society to take leadership in the promotion of the idea because no other national body existed which was interested in all branches of the industry. The Society is not seeking domination of the individual groups, nor is this desirable from any standpoint. It is available to you, however, as a central organization in contact with all of the ceramic industries and capable of being used as a clearing house for the efficient operation of such a federation as is proposed.

Booze pointed to the efforts of the Ohio Ceramic Industries Association:

> which has been able to obtain funds, equipment, and favorable legislation through the influence of massed attack. We are confident that the idea can be extended broadly with equally good results. Pressure can be exerted to secure larger government appropriations for ceramic work and equitable recognition obtained, which has been lacking in the past. Tariff, legislative and traffic matters can be dealt with better as a group than as individuals. Investigations on fundamental properties of ceramic materials and products which we are so badly in need of can be financed as a body, where individual organizations could not afford to do it, and a great deal of effort and expense can be saved by combining research programs.

and faience. It was the foremost pottery in the United States for nearly 50 years.

Mrs. Nichols was widowed in 1885 and subsequently married Bellamy Storer, a Cincinnati attorney destined for a diplomatic career. In 1890 Mrs. Storer divested her interests in Rookwood.

Josephine "Jo" Gitter was for more than 20 years an institution within the ranks of the Society. The daughter of a ceramic engineer from Zanesville, Ohio, in 1937 she was the first woman to graduate from The Ohio State University's ceramic engineering program. She was a self-employed manufacturer's representative and served as a vice president of the Society and as a Fellow. She was the first woman to serve on the Board of Trustees.

By many accounts Jo Gitter was the heart and soul of many of Society's best social events. From trap shoots to formal occasions, Jo Gitter brought a unique combination of frolic and formality. She was famous for maintaining her own suite at the annual meetings and was known for her parties.

"The real party never started until Jo Gitter got there," members recall. One remembered an occasion when Gitter's car broke down on the road to Bedford Springs. Alerted by a passerby, ceramists left the hotel in force and, attaching ropes to Gitter's car, lined up men in two columns to tow her car to the meeting site, where others were waiting to greet her like a latter-day Cleopatra.

"Her suite had flowers everywhere and a bathtub full of soft drinks," remembers Society staff member Alice Alexander. "There was a rack of ties behind the door, and no student came into her suite without wearing a tie. She was making gentlemen of them."

Recalls another recipient of Gitter's hospitality: "If you were a young student, you would act like young ladies and gentlemen, or you would not be welcome in her suite. She was a mother figure. She did a lot for hundreds of students by teaching them the professional way."

Another nearly legendary woman in Society recollection is Emily Van Schoick. Ross Purdy's assistant editor beginning in 1922, she worked as staff on Society publications for 25 years. Her career in ceramic literature spanned 40 years. Upon her retirement from the Society in 1947, she became the first librarian of the Scholes Library of Ceramics at the New York State College of Ceramics at Alfred University. She continued to consult on Society publications as a volunteer.

The New York State School of Clay-Working and Ceramics (now the New York State College of Ceramics at Alfred University) was the first to accept women students into the art and technology ceramics program, shown here in 1902. In the century's early years, women in ceramics mostly worked in pottery.

She edited the first edition of the Ceramic Glossary, *published in 1963. She was named a Fellow of the Society in 1936 and an Honorary Member in 1960.*

ACerS' women members work in many fields, but few as glamorous as space, the arena of Bonnie J. Dunbar, a NASA astronaut and Society member, who has served as one of three distinguished lecturers available through the Society as a speaker for meetings of sections, student branches and other Society groups.

Dunbar held a senior research engineer position with Rockwell International Space Division before joining NASA in 1978 as a payload officer and flight controller. She served as a guidance and navigation officer/flight controller for the Skylab *reentry mission in 1979 and was subsequently designated project officer/payload officer for the integration of several space shuttle payloads. She flew on a* Challenger *mission in 1992, taking with her a copy of the* Bulletin *and other ceramic mementos, now in the Ross C. Purdy Museum of Ceramics. Dunbar was a member of the crew that returned an astronaut from* Mir *space station in 1994 and was a crew member for the docking mission to* Mir *in 1998. She has also been a shuttle flight commentator for national television.*

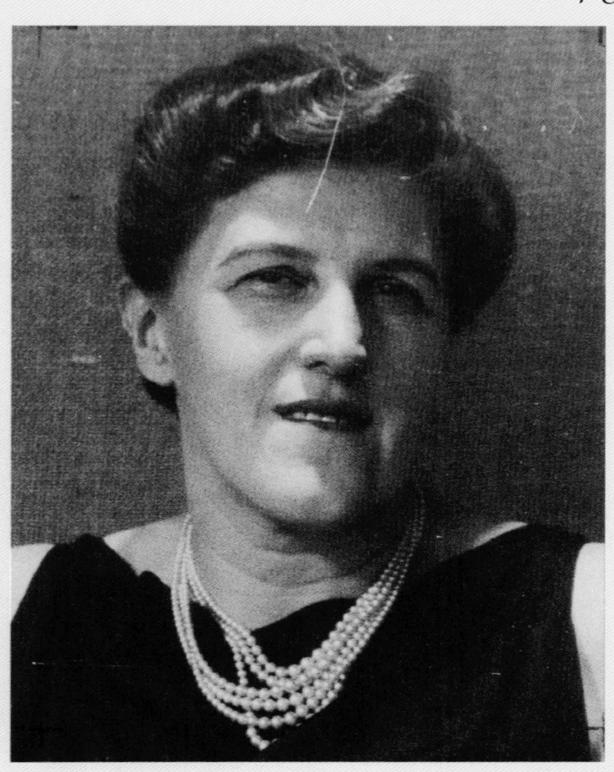
The legendary Josephine Gitter was, in many members' minds, the heart and soul of the Society's social side for more than 40 years.

Dunbar is a member of both the Engineering Ceramics Division and the National Institute of Ceramic Engineers. She was the editor of Materials Processing in Space, *one of the volumes in the Society's* Advances in Ceramics *book series.*

Women have served in Society leadership roles for decades, but nearly 100 years passed before the members elected a woman as president.

Carol M. Jantzen became the Society's first woman president in 1996. Jantzen, a Society Fellow, has served the Society in several capacities, including vice president-member development and vice president-at-large. She also has contributed to Society publications, serving as

The Society's grand old man, Edward Orton, Jr., who had just been nominated to return to the Society's presidency, weighed in on the topic. He noted the two schools of thought on where the Society should be going:

> One may be summarized in the idea that the Society is a scientific organization, whose business is to represent the technical aspects of ceramic science and technology and that this alone is a sufficient field for it to occupy.

Seventy years later, his words have a decidedly contemporary feel, as the question of the Society's appropriate and necessary sphere of influence still tries to balance science and the ceramics marketplace. Orton wrote in the *Journal*:

> Another group thinks that the Society must broaden out and make itself felt more quickly and definitely in the direction of making our present ceramic knowledge, which is constantly being increased by researches of the membership, more available for use by the various ceramic industries, in broadening their sales through the creation of new types of products and in meeting the competition which their products are constantly encountering from other nonceramic products.
>
> Of course such action cannot wholly be differentiated from promotion and sales work, and while promotion and sales matters are not properly within the direct purview of the Society, this group feels that it is the business of the technical men to greatly concern themselves with making their ceramic researches of more direct use to the commercial side of the business. These activities would necessarily include more or less expensive programs, including contacts with various groups through trade associations, the setting of standards and specifications, in order to represent ceramic products properly and to push their introduction and use.
>
> The Society will be compelled to make a choice. My personal belief is that the Society should not stand on the fence in this matter, and that it can and should, so far as the limitation of its budget permits, assist in interpreting its ceramic researches so as to make them more directly useful to sales organizations, and in promoting the business side of the ceramic industries.
>
> In other words, the decision between purely scientific and technical work FOR ITSELF and scientific and technical work AS A MEANS TO INDUSTRIAL EXPANSION is really the question at issue. My policy as president would be to have the Society develop in the latter line, as far as its budget can be made to go.

associate editor of the Journal *and contributing editor to* Phase Diagrams for Ceramists. *A member and past chair of the Nuclear and Environmental Technology Division, Jantzen is also a member of the Glass and Optical Materials Division and the Basic Science Division, the Ceramic Educational Council and the National Institute of Ceramic Engineers.*

Jantzen, a senior fellow scientist in the Glass Technology Group at Westinghouse Savannah River Co., Savannah River Technology Center, Aiken, South Carolina, holds a Ph.D. in materials science from the State University of New York at Stony Brook. A 1992 and 1993 recipient of the George Westinghouse Gold Signature Award, in 1998 she was nominated to the prestigious National Academy of Sciences.

Jantzen says, "I think that the attitude today is much improved for women in the Society and in ceramic industries in general than was the case when I joined the Society in 1970. There are many more women students entering the ceramic field than there were in my day. I remember taking a mining geology course and spending our field trips to mines in the parking lot with the bus driver because 'women were bad luck in a mine.' That won't be happening in a future that looks promising for ceramists regardless of gender." ▲

(Below) Emily Van Shoick (right center in front of Society staff) was presented with a silver cigarette case for 25 years of service in 1947 by Charles S. Pearce on behalf of the Board of Trustees. Her service to ceramic literature spanned four decades.

(Above) "Carol Jantzen asked me some years ago, 'How do you break into this old boys' club?' " recalled former Society president, William Prindle. "I said, 'Find something that needs doing, and do it.' " Jantzen (right) presided over the 1997 Annual Meeting as the first woman to serve the Society as president.

(Left) In the 20th century, more women have worked on the production side of ceramics than in technical or scientific capacities. In 1955, these women had factory jobs rolling milling media at Vesuvius McDanel in Beaver Falls, Pennsylvania.

(Left) More women than ever before are pursuing studies that will lead to careers involving ceramics. At the 1997 Annual Meeting, the James I. Mueller Outstanding Student Chapter award was presented to Angela J. Mercer by Virgil Irick, Jr., Keramos president.

It was a solid endorsement of Purdy's vision by Orton, a man who had sometimes seemed in Purdy's view to be the voice of the past, not the future of the Society. In the centennial year, this initiative of Purdy's endorsed by Orton, would still find supporters and opponents.

By 1930 The American Ceramic Society was bigger and more diverse than ever. It was so big and diverse that some members wondered if an overriding philosophy could be devised that would cover all of its aspects.

In his president's address for 1930, George A. Bole attempted to put it all into perspective:

By 1939, the Drakenfeld Colors Clayware Laboratory provided color matching, quality control and technical service assistance for ceramics customers using precision measurement, a far cry from the pinch-and-guess approach to ceramic manufacturing only 30 years before.

> The American Ceramic Society has grown so rapidly in the past few years and has come to include such a diversity of interests that we can scarcely classify ourselves more definitely than by saying that we are an association of people interested in the success of the ceramic industries.
>
> Some of us are chemists, others engineers, others artists, and still others are particularly interested in the field of management. Then there are those versatile chaps who solve production problems and indulge in sales activities. Some of us are in the laboratory, some in the plant, and others in the field.
>
> Service has become a major aspect of sales and the so-called trouble shooter is a permanent appendage to the sales department. The more technical aspects of sales are coming to demand the highest type of technical training.
>
> With such an array of diversified interests represented in our organization it is small wonder that we look at things differently.

Early attempts in the 1950s to host exhibitions had proved difficult and unprofitable, so for the first annual exposition, the Society turned over the arrangements to a professional show management organization. The exposition was a success and has continued to grow, attracting hundreds of exhibitors from the United States and around the world.

After years of consideration, the Society finally agreed to put concerted effort into expositions in the late 1960s.

> It is a challenge to our membership and a responsibility of our officers to blend so many interests into a coherent and purposeful organization.

Bole went on to detail examples of the various currents running through the Society:

- The proposal of a Science Division where papers of general scientific interest would be presented, coupled with the creation of the rank of Fellow of The American Ceramic Society to encourage and preserve scientific research.
- A special meeting of the Heavy Clay Products Division devoted to masonry construction problems, an example of the "pragmatic" bent of many members.
- Consideration of a creation of a Management Section and the holding of two management luncheons during the year.
- The raging debate over the appropriateness of the Society's participation in expositions. Many scientists argued that a technical society had no business sponsoring an activity so closely allied to sales.

Despite the forebodings of many, Bole noted that the Society would participate in an upcoming ceramics exposition in Cleveland:

> While some of us may be interested primarily in sales, some in production, others in research or in art, we are all at the last analysis interested in a product. Whether we are as a society to wash our hands of the product when it has been fabricated is the problem which we are now facing.

Indeed, it was a debate that would rage within the ranks for years. In editorial after editorial, Ross Purdy hammered away for a broadening of the Society's mandate, one in which the organization would serve not only a scientific and technological function within the industry, but promote ceramics in general.

It is highly likely that Bole (like many a Society president during this period) was echoing the thoughts of Purdy when he chided the membership, noting that:

> We have shown little tendency to produce new products or to find new uses for present ones. We have not utilized the skill of the artist or the genius of the research man to improve our ware. We have been imitators rather than originators. We have obeyed rather than created a demand for distinctive products. Only a few isolated instances can be pointed out where we have sensed a demand before it became articulate.

Part of the problem, as viewed by Purdy, Bole and others, was that for generations ceramic products were taken for granted. They were necessities of life, needed by every household and every business. But by the early 20th century nonceramic products were competing for the same markets.

The construction industry was a case in point. For years ceramic tiles, terra cotta trim and brick had been staples of the builder's trade. But now there were new, alternative construction materials available to architects and construction engineers — materials that were being aggressively marketed by their makers.

The ceramic industry, by comparison, could not even offer builders unbiased quantitative data about their products. The studies simply hadn't been done. And as a result, when building codes and zoning regulations were instituted around the country, many ceramic products weren't on the list of approved materials.

In 1929-1930 the Society began attacking the problem, sponsoring a structural materials conference in the Heavy Clay Products Division that would open dialogue between ceramic manufacturers, architects and construction engineers. Part of the program was establishing a standard unit size for each structural unit of material. And though the Society still shied away from outright promotion of ceramic materials, Bole called for ceramic trade associations to mount a new approach to merchandising their wares.

There also was talk of establishing a Ceramic Research Council to serve as a neutral fact-finding body sponsoring research and compiling test data. Such a group could speak with authority for all structural clay products.

A Structural Clay Products Research Foundation, an outgrowth of the National Brick Manufacturers' Research Foundation, was established by the Society in 1931. Created to produce materials such as handbooks, it was abolished due to financial difficulties in 1938.

The Karl Schwartzwalder-Professional Achievement in Ceramic Engineering Award for1989 was presented to Robert A. Rieger, TAM Ceramics, Inc., by ACerS President William H. Rhodes and Harold W. Stetson, president of the National Institute of Ceramic Engineers.

Awards are a form of celebration in all arenas of the Society. Here Robert B. Oberst, president of the National Institute of Ceramic Engineers (NICE) in 1991 presents Thomas D. McGee with the Alfred Frederick Greaves-Walker Award at the NICE/Keramos Luncheon.

(Below) Della Roy, Orton Memorial Lecture Committee, presented J. Derek Birchall a plaque after his Orton Memorial Lecture in 1993.

AWARDS AND HONORS

One of the Society's obligations and privileges has always been to honor exceptional accomplishments in the field of ceramics. One of the first acts of the newly formed Society at its first meeting in 1899 was to recognize Mrs. Bellamy Storer by electing her as the first Honorary Member in recognition that she was the first to produce a distinctive type of American art pottery ware. Over time, more than 50 awards have been designated by the Society, its divisions and its sections. Many are no longer in existence, having become irrelevant, unfunded or simply forgotten, but many have stood the test of time and are still a very important part of the Society's legacy.

The grade of Honorary Member, first given to Mrs. Storer in 1899, was dropped some time later, but reestablished in 1982, and the selection process is virtually the same as it was in the early 1900s. Recipients are chosen annually from among those nonmembers of professional eminence nominated for their achievements in the ceramic arts or sciences.

Initially designated as Honorary Life Membership, the honor of Distinguished Life Member was established to recognize current members of the Society of professional eminence elected for their achievements in the ceramic arts or sciences or their service to the Society. The honor was renamed in 1982 when the grade of Honorary Member was reintroduced. A maximum of three Distinguished Life Members may be elected annually.

The grade of Fellow was established in 1930. In President Orton's own words, "The work of the Fellowship is to build an esprit de corps and incite the young, who are beginning, to put forth their best efforts to real achievement." Initially, the first One Hundred Fellows were chosen by ballot and invited to meet at the Society offices on December 27, 1930. The first Fellows induction ceremony was held during the 1931 annual meeting with 153 inductees. The nomination process agreed upon in 1930 is virtually the same as that followed today. Fellows of the Society must be at least

Reaching the rank of Fellow in the Society is an honor attained by fewer than 5 percent of the members in 100 years. Here, Thomas F. Root received the Fellow Award in 1991 from President Dennis W. Readey.

Lectures represent a special kind of recognition that provide an honor for the selected lecturer and an opportunity for information to the membership. In 1991, Arthur H. Heuer presented the Orton Memorial Lecture.

Two long-time members of the Society celebrated the 80th birthday of Dong Sheng Yan, February 10, 1998. Yan (right) and David Kingery are each Distinguished Life Members of The American Ceramic Society. In Kingery's honor, the Kingery Award was given for the first time at the 1998 Annual Meeting.

(Right) Honors and awards create opportunities to recognize individuals and organizations for many kinds of support to the Society or achievements in ceramics. Corporate participation is heralded in a variety of ways, including the Honorary Membership Award, presented here in 1991 to James R. Houghton (Corning Inc.) by ACerS President Robert J. Eagan.

"An engineer's handbook of structural clay products containing all the known data required by architects and builders would do more than any one thing to stimulate a lagging confidence in fired clay structural materials," Bole wrote. "Such a handbook should be put out by a technical society and not by a trade association."

Meanwhile, Ross Purdy again became involved in a war of wills with the Board over Society involvement in ceramic exhibitions.

Purdy was convinced that The American Ceramic Society should play a role in the Century of Progress Exposition at the 1933 World's Fair in Chicago. He envisioned a structure built and decorated with ceramic materials in which the industries collectively could show the century of progress made in ceramic manufacturing and the vast variety, the high quality and the utilitarian value of American ceramic products.

The project, he wrote, would be designed:

> to make Americans more ceramically conscious; conscious that in America, clay, glass and glass enameled ware are being produced which are superior in several respects and inferior in no respect to ware produced elsewhere and certainly having far greater serviceability than any of the nonceramic products designed to serve like purposes.

The Board, though, viewed the project as too expensive; some members thought the Society had no business putting on expositions, which they viewed as little more than glamorized merchandising. Purdy was ordered to cease all activities regarding the Century of Progress Exposition. He defiantly answered that he'd continue to work on the project in his spare time.

"It makes me blush with shame when The American Ceramic Society is called upon to look into a matter that it has to limit its activity to merely collecting the existing information and has to leave it to other associations to organize and to prosecute the investigations," Purdy railed in print. ". . . an American Ceramic Society that is strictly academic would be almost valueless. This Society must be aggressively participating in the studies of products, and the uses of products if it is going to be of any value to ceramic industries."

Eventually, Purdy won the backing of several influential members, including once again, Orton.

In his presidential address of April 1931, Edward Orton, Jr. exhorted the members:

> I think the Society should definitely set itself to serve more directly the ceramic industries on their business side. They think of us now as a group of tech-

35 years old, have been members of the Society for at least five years continuously and have made outstanding contributions to the ceramic arts and sciences, through broad and productive scholarship in ceramic science and technology, by conspicuous achievement in ceramic industry, or by outstanding service to the Society.

In addition to special grades of membership, awards and lecture series have continually been added to the Society's portfolio of honors. The first lecture and the first award were, appropriately, created to honor the two most influential men in the Society's early history.

The first Society lecture series was established to honor Edward Orton, Jr., who died in February 1932. The first presentation in the Edward Orton, Jr. Memorial Lecture Series was made in 1933 during that year's annual meeting by E.W. Washburn, who spoke on "The Phase Rule in Ceramics." The honor of presenting this highly regarded and well-attended lecture is given by committee to a speaker nationally recognized in a field related to the interests of the membership of the Society.

The Ross Coffin Purdy Award was established in 1946 to honor technical and scientific advancement under the name of the long-serving and influential Society general secretary. The award is given today to the author or authors who, in the judgment of the committee, made the most valuable contribution to ceramic technical literature during the calendar year prior to the selection.

Society officers have historically expressed a concern that the general membership was not adequately made aware of the award opportunities offered. Today, thanks to modern technology, members need only access the Society's World Wide Web page to get a complete listing.

Many of the divisions and sections of the Society administer their own awards and honors programs as well.

GUARDIANS OF THE SOCIETY

The class of Fellow of The American Ceramic Society was created in 1930 to recognize members who have made exceptional contributions to the ceramic arts and sciences.

To a great extent, it was a reaction to the "watering down," in the minds of some members, of the scientific and technical missions of the Society. Increasingly, persons with only a cursory interest in pure research were joining the Society, causing founder Edward Orton, Jr. to observe that, "We are the only important society in the country which makes admission extremely easy, and which at the same time gives no recognition to professional skill or scientific attainment."

Orton noted that in the Society's early years, associate

"As history is recorded, some persons do not get the credit they deserve. . ." began a student speaking contest entry at the 78th Annual Meeting in 1976. The subject of the address was Samuel Geijsbeek (1870-1943), one of the founders of the Society, and a ceramist whose technical skills were widely admired. A Dutch chemist who worked with Hermann Seger, Geisjbeek came to the U.S. to enroll in Orton's new ceramics program at The Ohio State University, and it was Geijsbeek whose signature appeared first on the letter inviting a select group to meet as the "Association of Ceramic Chemists," a group that became The American Ceramic Society. He went west, created the first whiteware west of the Mississippi in 1902 in Colorado, then went on to Seattle, where he became the first West Coast member of the Society. His obituary in the Bulletin *recalled the generosity of his intellect at a time when glazes and other technical information were secret: "He could not believe that any person would take advantage of his liberality: He would not protect his own interests but would give freely and wholly of his talents and trust."*

Albert V. Bleininger (1873-1946), an immigrant ceramist born in Germany, was a Society Charter Member, friend and colleague of Geisjbeek's. Bleininger is chiefly remembered for translation of Hermann Seger's collected works; for U.S. ceramists it was a groundbreaking scientific document and its publication was one of the early achievements of the Society.

The Society's offices at 2525 North High Street were decorated throughout with portraits of Charter Members, Fellows, officers, notable ceramists and others in the pantheon of the Society.

members viewed themselves as probationary members.

> They had to write papers, to make discoveries, work on Society committees or become noted as plant managers or executives. In short, they had to ARRIVE in some way and in some part of the field; otherwise they could not become active members.
>
> Now with this spur to personal ambition gone, many members who could and who should do a great deal for the Society do not do anything, because they can get the nominal fruits of membership without doing any work whatever.

In the beginning the Fellows were seen by Orton as a way of encouraging and rewarding pure scientific research. To be eligible for election as a Fellow, an individual must have been enrolled as an Active Member for not less than five years, must have attained recognition and prominence in the Society by high professional ideals, sustained interest in and devotion to the objects of the Society, and demonstrated broad and productive scholarship in Ceramic Science or achievement in some branch of the Ceramic Arts.

Today, Fellows are chosen from among those 35 years of age or more, who have been members of the Society for five continuous years, and who have made outstanding contributions to the ceramic arts or sciences through broad and productive scholarship in ceramic science and technology, by conspicuous achievement in ceramic industry, or by outstanding service to the Society.

The Fellowship, Orton said, was:

> the spiritual guardian of the Society, the preserver of its ideals, the stimulator of its pride, the radiant force of its inspiration, whose influence shall lead us always to higher levels of achievement in the future.

Through 1998, 1,537 members have been selected for this honor. ▲

> nical men who read papers to each other. They do not visualize in us a powerful and resourceful ally to bring in more business or to make their present business more profitable. . .
>
> Ceramics industries need aggressive leadership of national character to represent their products. They need representation in the places where standards are set and specifications are drawn. They need representation in government departments, where promotion plans are made and public questions are answered. They need sustained and intelligently developed publicity through which the people may be forced to learn the merits of what is produced, and the value of the services their products can render.

Noting that for too long the Society had been "passive in character," Orton then endorsed Society participation in the 1933 Chicago World's Fair, the Century of Progress.

> The ceramics industries must participate . . . to absent themselves from it would be simply suicidal. To participate singly as competitors, without any effort to set themselves and their industry forward as a whole, would be almost as bad as to refuse to participate at all. What is needed is a ceramic building, built of ceramic products, to house educational exhibits and working models of what we have done in 100 years to serve mankind. The Society cannot supply funds, but it can supply everything else: intelligence, energy, skill, incentive and leadership . . .
>
> Our opportunity is here. It remains to be seen whether we are big enough intellectually to see it, and big enough in leadership and spirit to grasp it.

But it didn't happen. The Board decided the money just wasn't there.

Over the years, The American Ceramic Society participated in four ceramic exhibits — in 1929, 1931, 1948 and 1951. Of these none were financial successes and two were near-disasters. After the 1951 show the Society sought the services of a company specializing in managing such events. This firm surveyed the situation and politely declined to get involved.

(Below) Early in this century, a ceramic lab looked something like an apothecary shop, as men like Fortunatus Quartus Mason mixed their compounds from a blend of experience and science.

Students at the New York State School of Clay-Working and Ceramics around 1900 made dry-press ceramic tiles. Until World War II, the ceramics industry in America was largely clay-based.

"Since then the trade show idea has lain dormant, roused only gently by members unfamiliar with its unpleasant past," the *Bulletin* wryly noted in the mid-1950s. It was not revived successfully until the late 1960s.

While Purdy envisioned ceramic exhibits that would inform the public of the "utility of ceramics," later, more successful exhibitions were more business-to-business than promotion oriented. From Purdy's initial enthusiasm, a more clearly defined use of exhibitions evolved, providing members with better information and opportunities for productive relationships.

THE CLASSES

If Purdy's dream of The American Ceramic Society as an all-encompassing umbrella organization for the ceramic industry had not yet been realized, the Society was nevertheless taking steps to bolster its reputation and vitality.

Two of its most important efforts were the Institute of Ceramic Engineers, today called the National Institute of Ceramic Engineers, and the Ceramic Education Council, both created in 1938. These organizations constituted the two classes of the Society, and members were allowed to belong to them in addition to any divisional affiliation they may have had. The 1938 Constitution provided that the divisions divide the membership by product; the classes by training or interest of the individuals. Today, these distinctions are blurred, and both divisions and classes attract members with a broad range of interests.

"Ceramic engineers find themselves in a position where, in order to maintain their status and in order to obtain a status among other engineers, it is necessary for them to have some sort of an engineering organization," argued Arthur F. Greaves-Walker, a prime mover behind the creation of the Institute. He insisted that membership in the Society — a technical organization rather than an engineering group — was inadequate if ceramic engineers were to hold their own with their peers.

Clay production methods that dated back to the 18th century innovations of Josiah Wedgwood were still in use at the turn of the 20th century, but ceramics processes and production changed dramatically over the next 50 years.

Even before the Institute was formally organized, some members of the Society had represented themselves to other engineers as members of the Institute of Ceramic Engineers of The American Ceramic Society, "and it gives them a status they haven't had and which is absolutely necessary," according to Greaves-Walker.

The Institute was designed to promote the professional status of ceramic engineering, high standards of ceramic education and high ethical engineering standards and practices. It has also offered an employment service and assisted its members to become registered professional engineers.

The role of the Institute within the framework of the Society, according to its organizers,

> is to engage in all activities of primary concern to the ceramic engineer. It is our responsibility to participate in the accreditation of the schools where ceramic engineering is taught; to provide opportunities for the continued education of ceramic engineers by the dissemination of ceramic engineering knowledge, to establish opportunities equal to those of other engineers for professional registration, to instill in ceramic engineers worthy standards of professional ethics and to encourage professional achievement in ceramic engineering.

The Institute's creation was just what the "practical" men in the Society had waited for — an avenue for discussion of ceramics from their point of view, rather than by the scientists whose outlook had dominated the proceedings for 40 years. And, not unexpectedly, the marriage between the Society and the Institute has often been rocky.

Just how profound the conflict between science and engineering has been may be difficult for the layman to grasp. One attempt to analyze the situation was made by J. Herbert Hollomon of the U.S. Department of Commerce in his Orton Memorial Lecture of May 9, 1966. Among the points Hollomon made:

> The job of technology is to provide the understanding, the information and the background to bring the fruits of science to the practical benefit of civilization and society.
>
> Science is an invigorating, challenging, curiosity-seeking activity. Technology is the application of knowledge to practical purposes.
>
> . . . It often takes 25 years for the science that is available to be put to use. The basic reason for this significant lag is that the science that most engineers and technologists learn is current when they learn it in school, but it is outdated before they become productive, working in industry or in government.

Decorated pots of the kind the New York students made were prized in the Society's 2525 North High Street collection.

Men and women in the New York State School of Clay-Working and Ceramics (now the New York State College of Ceramics at Alfred University) did finishing work about 1928. By that time, art and design had begun to separate from engineering and science within the College. In the background, the male student is Paul Vickers Gardner, who studied under Charles Fergus Binns, worked with Frederick Carder at Steuben Glass Co. and was the first curator of ceramics at the Smithsonian Institution.

> . . . The scientist is intent upon knowledge for its own sake. The technologist invents something that will work and be useful. Frequently, the new products and services which are delivered to a society involve things which are not understood by the people who invented and developed them, and the understanding only comes after the development of the product or service.

Today the debate continues, still a vital part of the intellectual and professional interaction that enlivens Society meetings and correspondence.

Much in the way the Institute serves ceramic engineers, the Ceramic Education Council stimulates, promotes and improves ceramic education.

It provides a national forum for the discussion of issues pertinent to ceramic education, curricula and institutional affairs. It encourages interaction between ceramic educators, provides recruiting assistance to the ceramic schools and supports the National Institute of Ceramic Engineers in matters of ceramic education and accreditation.

WORLD WAR II

"It required a world war to bring to ceramic engineers the full realization that although their branch of engineering is one of the oldest, it is one of the least known," Arthur F. Greaves-Walker wrote.

In contrast to the Great War, in which ceramists were called upon by their country to develop new and better materials for the effort, World War II found members of the Society adrift, largely overlooked by the nation's mobilization effort.

"It has been somewhat of a shock to many to find that the Selective Service System, the Army and the Navy have refused to recognize that ceramic engineering, except in a few cases, has any place in the war effort," Greaves-Walker observed.

Without favored status, the ceramic industry was quickly decimated by the draft and by a general lack of government interest in its capabilities and potential.

For example, while the Selective Service Administration listed professional and technical engineers as among the critical occupations that might be draft exempt, Selective Service memos did not mention ceramic engineers.

Unless an official ruling was made specifically including ceramic engineering as vital to the war effort, Society leaders warned, the ceramic schools were likely to find their departments effectively abolished for lack of students.

Indeed, that is exactly what happened. The *Bulletin* reported in 1943 that in the previous year the ceramic engineering departments of U.S. universities had 202 requests for graduates, of which they were able to supply only 53. Every able-bodied young man of college age, it seemed, was in uniform.

It was a period of belt-tightening for the Society — it is said that Ross Purdy paid his assistant's salary out of his own pocket.

In 1944 the Board voted to create the position of assistant secretary, with the duties of serving as understudy to Ross Purdy, to promote increased interest in the local sections and the student branches and to supplement the other activities now being performed by the general secretary. Charles Sidney Pearce was hired for the job and two years later, upon Purdy's retirement, would succeed that remarkable individual.

Charles Pearce shared Ross Purdy's fervor for the national and international acceptance of The American Ceramic Society as the scientific spokesman for the ceramic industry. But like his predecessor, he, too, was frustrated by the vast possibilities of the Society and the constraints — most of them financial — that kept them from being realized.

The Society, Pearce lamented in 1950 with not a little hyperbole,

> is still a relatively unknown organization with offices in a depreciated loft building at the edge of Columbus. The managements of the progressive and wealthy corporations in the industry still view The American Ceramic Society with little concern, if they even know of its existence, and the managements of the national professional societies look on it as of little consequence.

And, he groused, "The development of an outward-looking and forward-looking policy for the Society has not been accomplished."

That was a lament that would be heard with regularity for another 20 years.

The post-war years were, for the most part, good for the Society. Increased federal funding for research and development — exemplified by the creation of the National Science Foundation and the Office of Naval Research — provided new opportunities in the ceramics field. By 1951 the annual budget had increased to $150,000 — about three times that of the pre-war years. An emphasis on building the Society led to a Double the Membership Campaign. The membership didn't double, but it increased, and perhaps, more importantly, the effort encouraged recognition of the need for future emphasis on member recruitment and retention.

In 1955, entertainment at the Annual Meeting in Cincinnati packed the grand ballroom of the Netherlands Hotel.

The Committee on Ceramic Education was organized with representatives from six divisions, two from schools and one from the Institute of Ceramic Engineers. The Committee's mandate was to launch a program to promote enrollment, which had been dropping in ceramic schools. The group's first effort was to create a brochure to introduce high school students to careers in ceramics. The item proved to be a popular perennial, and over the years went through several reprints and redesigns.

The most positive step of the 1950s was construction of the Society's new home. In 1954 the Society moved into its own built-from-scratch headquarters building in Columbus, Ohio. It was a sign of affluence on the part of the Society; of even more importance, it was an indication of permanence. The American Ceramic Society was here to stay.

The era was not without controversy. In particular, there was the ticklish situation of some of the divisions, which were feeling so independent that they had elected officers who, in many cases, were not even members of the Society. This served as a warning to the leadership that it was unwise to take a laissez-faire approach to governing. Rules were rules.

"Presently the Society consists of 10 separate groupings called classes and divisions, which are operated much as 10 separate societies," observed one Society officer. "This multiplicity of organizations creates many of the problems of 10 separate associations and entails much cost."

There were also calls for the Institute of Ceramic Engineers (now the National Institute of Ceramic Engineers) to break away and become a separate entity. It was the old engineers-vs.-scientists scenario being played out yet again. Nothing came of the secession talk.

Another burning issue of the 1950s was just how far the Society should go in publicizing and promoting ceramics on a national level. The question — the same one Purdy grappled with throughout his career — was raised and debated numerous times, but always the final determination was that the Society had no business in that business. Yes, the Board always concluded, the Society is the only link between the various branches of the ceramic industry, but it should not wander too far beyond the technical realm.

"There are 25 or more trade associations in the ceramic industry . . . it is the individual responsibility and opportunity of these 25 ceramic trade groups to carry on publicity for their own benefit," the Board declared. "It certainly would be impractical and unwise for The American Ceramic Society to attempt a program which would be spread so thin as to cover the hundreds of ceramic products made by industry."

There was some compromise though. In 1953 the Board created a publicity committee for the Society and for the ceramic industry in general.

In 1958 membership passed 7,000, but the treasurer's report that year warned that "in the face of ever-spiraling costs, disturbing shadows begin to cloud the picture." Poor advertising support in the *Bulletin* greatly increased competition from for-profit trade magazines and a whopping 35 percent increase in the number of pages in that year's *Journal* were cited as major concerns.

Indeed, the Society entered the 1960s with a budget deficit and a commitment to fiscal belt-tightening.

When a survey revealed that 40 percent of those attending Society meetings were not members, it was the impetus for yet another membership drive. It worked. Membership and income grew. An effort to build a much-needed expansion to the headquarters building raised $100,000 from individuals and corporate donors, and the construction was completed in 1967.

In 1961 Vice President Eugene C. Clemens, citing "competition between nations of the world for supremacy" urged Society members to help the United States retain its competitive edge and economic advantage over other nations. In the area of structural clay products, his suggestions included attracting capable engineers and scientists to enhance operations and research, encouraging ceramic schools to train more men for the field, instituting more research at the plant or manufacturing level, and elevating the division to the same level as others within the Society. Clemens believed that in the rapidly changing organizational structure of the Society, these steps would be necessary to prevent the structural clay products division from being swallowed up during impending reorganization.

Indeed, the ceramics industry at large had undergone a shift in the years after World War II. Before the war, ceramics was largely a clay-based industry. In the second half of the 20th century, a wealth of new materials created different opportunities for ceramic applications and engineering.

Suggestions for competitive strategies appeared reg-

In the second half of the 20th century, ceramics has shifted from a clay-based industry (right, center) to an industry that solves problems through ceramic processes in many kinds of applications and materials. Working with allied technologies, ceramic engineers found an increasingly open field of innovation in the 1980s when funding for research and production was plentiful.

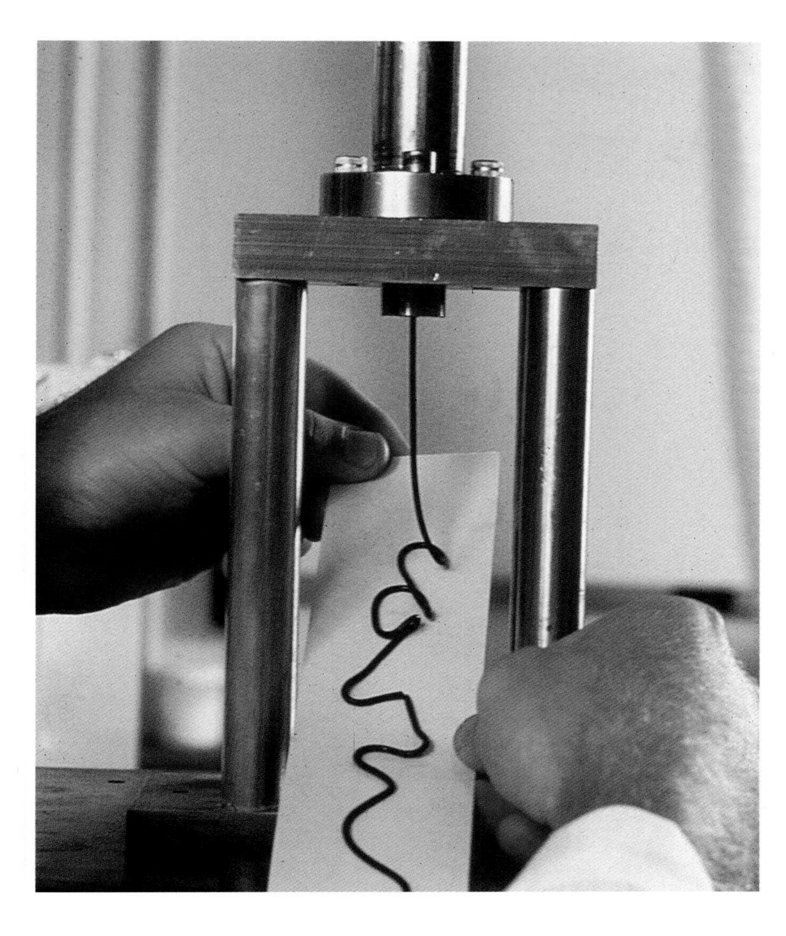

ularly in the *Bulletin*. In May 1962 the *Bulletin* reprinted comments from a Society Symposium presentation by U.S. Department of Commerce representative Edward R. Killam. In it Killam offered his perspective on the questions the ceramic industry should ask itself, such as, "Are you getting your share of the expanding domestic market?" and, "Are you planning for the regional and population shifts being experienced in this nation?" Expanding research and development efforts to meet foreign competition and taking full advantage of government aid to expand export were other suggestions Killam made for sustaining the "productivity of our expanding economy."

By 1965 the Society had a paid staff of nearly 30 persons — triple the number of 20 years earlier. In 1966 computer equipment was installed to handle membership records and accounting chores.

And finally, 30 years after Ross Purdy began complaining about the impossibility of administering the Society while editing its publications, the duties of general secretary and editor were split. A new position — Editor of Publications — was created. General Secretary Pearce became the Society's publisher and maintained overall responsibility for books and periodicals, but was freed of much of the hands-on editing.

As Pearce was fond of saying, "An editor lives by contention and a secretary lives by compromise. The combination of these two functions leaves a journalistic zero."

In 1967, a long-range planning committee was created. In the decades since, planning has addressed both recurring challenges and new opportunities. Among the committee's first recommendations was that the Society should actively participate and take a leadership role in cooperative activities with other scientific, technical and professional societies in the areas of education, technical information service, professional characterization, government liaison, international commissions and councils.

Among the other interests mandated as the province of the committee were: size, growth and character of the membership; trends in the meetings and their programs; trends in the volume and nature of technical articles, special publications and abstracts; the Society structure as represented by classes and divisions; activities and procedures of the sections; administration and staff requirements; member services, including continuing education and technical information services; public relations; ceramic engineering student enrollment; relations with other technical societies in the United States and abroad; estimated future operating costs and how they may best be financed.

In short, just about everything that was on anybody's mind. But as broad as the committee's charge was, it was still a modest assessment of an increasingly complex organization.

In 1968 another of Ross Purdy's and Edward Orton, Jr.'s dreams finally came to fruition — although in a somewhat different form than they probably imagined: the Ceramics Exposition was approved. The first such Exposition was held at the Society's annual meeting in Washington, D.C., in 1969. It wasn't the public education extravaganza Purdy had advocated for the World's Fair, but it did showcase ceramic innovations and production. Nearly 100 companies reserved booth space and displayed raw materials, products and services. Show World, Inc., a New York firm, was retained as consultant and exposition manager for the three-day event at the Sheraton-Park Hotel.

At an Annual Meeting in Atlantic City in 1947, Loran O'Bannon, L.I. Shaw and Winston Duckworth strolled the famed Boardwalk.

Executive Director Frank Reid, who succeeded Pearce in 1963, and during whose term the traditional title of general secretary became the more contemporary "executive director," enthused:

> An Exposition will prove a valuable educational service to our members. If favorably received, it will enhance the Society's public image, become a continuing source of new members and provide additional revenue which is badly needed to underwrite valuable nonrevenue-producing services of the Society.
>
> The Exposition program has grown to include exhibits not only at the annual meeting but division meetings and workshops as well.

Indeed, the Expositions have proved valuable, providing a service to members and delegates, producing income sufficient for the Society to offset its expenses and creating some public exposure.

After World War II, with the advent of speedier communications and easier travel, the international ceramics community was more connected than before. Society members played important roles around the world, including the installation of Andrew I. Andrews as president (seated at table) of the Third International Congress on Vitreous Enameling in Venice, Italy, in 1961.

At Pennsylvania State University in 1956 a student used a thermal expansion recorder.

In 1970 the Society's total income was approximately $770,000. Of that, roughly 35 percent came from dues and subscriptions, 30 percent came from advertising, and 15 percent came from meeting revenue. The balance came from other miscellaneous sources.

In 1970 the establishment of the Ceramic Endowment Fund enabled friends of the Society to make contributions to be permanently set aside to foster particular aspects of the Society's activities.

President J. Earl Frazier noted:

> It is the firm purpose of the Board of Trustees that the Ceramic Endowment Fund not be used to compete for funds which currently are and in the future should be available to the Society in the course of its present operations. . . . It is hoped that the existence of a ready vehicle within The American Ceramic Society for the perpetuation of particular gifts designated for the support of specific activities or purposes will encourage a different kind of giving, one for the lack of which the Society has been suffering.

A long-time project of former President William Payne, the endowment efforts of the Society have resulted in the recent establishment of the Ceramic Foundation. The Foundation will use donor gifts to promote the art and science of ceramics for the use and benefit of the public.

Flush with success, the Society went into the real estate business, taking out a bank loan and constructing the Ceramic Park Building just behind the headquarters. Offices were rented to a variety of concerns. It was

A GRAVE CONCERN

In 1953 the Society squared off against the U.S. Postal Service in a dispute over the dissemination of Soviet technical material in this country.

This was, of course, at the height of the Cold War, and the solicitor general of the post office instructed the postmaster at Cleveland to seize certain technical literature sent from the U.S.S.R. to Society headquarters on the grounds that it contained Communist propaganda.

This action outraged members of the Board, who adopted the following resolution:

> *The Trustees of The American Ceramic Society, believing that the availability of technical information from all possible sources is of vital importance in maintaining the technological supremacy of the United States, hereby*
>
> *Resolve, that any attempts by the Post Office Department, or any other agency of government, to restrict the free flow of technical information into this country, whether or not such information contains matter that may be construed as communistic propaganda, be viewed with grave concern.*

Copies of the letter were sent to the postmaster general and to the chairman of the Foreign Affairs Committee of the U.S. House of Representatives.

Noted Secretary Charles S. Pearce: "It is deemed that members of The American Ceramic Society are of sufficient technical competence to understand this material, and of sufficient patriotic competence to reject subversive propaganda." ▲

hoped, Reid said, that the office building "will add financial stability in the years ahead."

Ironically, financial stability was the last thing the Society was about to realize. In 1978 the new treasurer, James Mueller, discovered that the Society faced bankruptcy by the end of that summer. Basically, the Society had been borrowing money to pay its bills.

For years it had been the custom of the treasurer to simply sign over to the general secretary 12 blank checks, one for each month. This was a system that invited disaster, as it gave the Board little or no oversight function into the spending of Society funds.

As times have changed, the Divisions have changed to meet them. For example, in December 1961, the Enamel Division changed its name to the Ceramic-Metal Systems Division after considerable discussion of the expanding scope of interests of members. Division Chairman Loran S. O'Bannon of Battelle Memorial Institute (right) and R.S. Sheldon, Division Vice Chairman, listed some name change choices at the Division's first annual fall meeting that year.

Fiscal housecleaning ensued.

Staff member Gary Panek found a way to defer a tax payment for about six months, allowing the Society to squeak through. But it was touch and go — older members recall counting the receipts from the Refractories fall meeting that year, desperate to come up with just enough money to pay the Society's bills through December.

The crisis led to many long and acrimonious meetings. Bob Beals, who held the treasurer's slot after Mueller, spent many weekends in Columbus working on finances. It took nearly three years to turn the financial situation around.

As the decade closed, the Board voted to approve multiple division memberships. Each Society member could belong to a maximum of three divisions. This option allowed members "to combine their interest in science and engineering and between the divisional interests such as glass and nuclear, cements and refractories, and refractories and glass. The recommendation provides for a furtherance of cross-fertilization amongst the divisions and for nurturing interdivisional interests."

The 1980s were boom years for the Society and for ceramics in general. Reaganomics was good for the industry and the

(Right, bottom) Flush with success in the mid-1970s, the Society developed an office building behind the Ceramic Drive headquarters and rented space to a variety of concerns. When the staff needed more space, the office building was considered but it had been built across the ravine (foreground) from the main building, so reasonable integration of staff would have been impossible. By the early 1980s, it was time to move.

Herbert Insley, W. E. S. Turner of Sheffield, England, and Charles S. Pearce (left to right) conferred on publications in 1957. Insley served in many capacities in the Society, including editor of Society publications from 1954 to 1972. He is most widely remembered for his work on Phase Diagrams for Ceramists, *which, in the opinion of many people, has been one of the most important efforts of the Society's first century.*

The doors to Binns-Merrill Hall at the New York State College of Ceramics at Alfred University have welcomed several generations of ceramic engineers and are a visual reminder of the doors to opportunity represented by ceramics throughout this century.

federal government lavished money on ceramic research, which, in the minds of many, was the wave of the future. The possibilities of cable optic fiber and the development of ceramic heat shields for use on the U.S. space shuttles suddenly made ceramics a glamour field.

A survey in 1982 revealed that among research and development professionals, ceramists recorded the greatest number of pay increases, with nearly 90 percent of all ceramists earning salary increases of at least 6 percent.

Ceramics were seen as the solution to numerous military and economic problems; suddenly the industry was flush. For the first time the Society could afford to foot the bill for air fare so that the president of the Society could attend important meetings. In years past the president paid his own way — indeed, one of the criteria for holding that position was a willingness and ability to finance one's own travel to and from meetings.

Capitalizing on this federal appetite for all things ceramic, the Society hired Hamilton Associates, which also represented the American Institute of Chemical Engineers, as its Washington representative.

In an address to the 1980 fall meeting of the Materials & Equipment and Whitewares Divisions, the new lobbyists explained the Society's role in Washington:

> Because of its tax status, the Society cannot lobby on behalf of legislative or regulatory proposals that will benefit its members. But The American Ceramic Society can be effective — and in the long run benefit both its members and the public — by providing technical data, technical review and consulting to Congressional staffs and regulatory agency personnel.

This was also the decade when the Society was recognized as truly international. It wasn't so much that the Society had a huge foreign membership, but rather that the *Journal* was the preferred publication route of the world's finest ceramic scientists.

As early as 1967, the Long Range Planning Committee recommended Society participation in international commissions and councils and possible abstracts. This growing advocacy for international connections resulted in the creation of the International Affairs Committee, which helped the Society maintain a prominent role in an increasingly global field.

Shortly after it was established in 1979, the International Affairs Committee began to help the ACerS realize its vision of building international stature and forging new relationships with ceramists overseas. Xenophobic attitudes generated by the Cold War and Common Market alliances had been assuaged, and worldwide circulation had been producing significant revenue since the 1950s. The Society logically believed that the potential for growth existed in the international scientific community, and the committee's coordinating efforts with similar organizations overseas seemed an obvious opportunity to gain knowledge and exposure in other countries.

By initiating technical exchanges between scientists and engineers in the area of ceramics, the committee hoped to encourage familiarity with other countries' ceramic techniques and sciences. Society publications would offer some opportunities for information exchanges; delegates visiting other countries also could take part in paper presentations, group discussions, seminars, individual conversations and correspondence. Most importantly, these visits would help ceramists develop personal relationships with their international peers, allowing them access to long-term information about how demographics and economy affect the ceramic industry.

In 1980, the Society's inaugural international visit took a 17-member delegation led by Malcolm McLaren and including President William Prindle and the executive committee to the People's Republic of China. The Chinese delegation, headed by Xu Zhuo-ran, organized lectures by Society members that were attended by Chinese ceramists from across the country. Interpreters translated the lectures for the Chinese scientists and engineers.

Americans visited a number of factories, learning much about how the Cultural Revolution had immobilized universities and stopped advancements in engineering and science. By the time of the delegation's visit, university graduating classes were finally returning to their pre-revolution levels.

One of the highlights of the trip was a visit to one of China's great archeological treasures, the tomb of the Emperor Ch'in Shih Huang Ti and his 7,000-man ceramic army, created nearly 2,200 years ago. There was much speculation on how a technologically primitive society could create such an overwhelming artifact. The visit gave the delegation an opportunity to see ceramics in a historic context unlike any they could experience at home.

Even as ceramic processes and materials have taken new methods and forms, ceramics of all kinds still have in common a need for heat.

By the early 1980s, significant research by Society members was an important scientific exchange in the exploding arena of international communications and commerce. To distinguish the organization's name in arenas (such as chemistry) where research efforts were publicized, the Board of Trustees settled on the acronym ACerS.

After World War II, international visits broadened from the European exchanges of earlier years and included guests such as the ceramists from Israel, shown with General Secretary Frank Reid in 1956.

Hans Hausner (podium) presided over the Fourth CIMTEC meeting in Saint-Vincent, Italy, in 1979.

Thanks to the Chinese hosts' warm reception and the well-rounded activities they had planned, the visit — which included the first Society Executive Committee meeting outside of the United States — was deemed a success. The information shared and the mutual goodwill conveyed established a pattern for similar visits and better communication with societies abroad.

Also in 1980, the Ross C. Purdy Museum of Ceramics was established in the Society's headquarters. The avowed purpose of the museum — based on the collection assembled by Purdy during his years as general secretary and left to the Society — was to

> collect, preserve, display and research historical material on objects of ceramic manufacturing, production and use, and of the individuals and firms associated with the development of the ceramic industry and ceramic science in America; to own and operate a museum of contemporary industrial or historically significant ceramic objects and materials. To interpret the history of the manufacture of ceramic items, including the progress of the arts and sciences concerning the production of ceramic objects and products; to publish material concerning the museum and its objects, collections or displays; to conduct educational programs related to the ceramic museum; and to do anything that is worthwhile to achieve these objectives.

By the beginning of the 1980s, the Society's position in scientific circles — and its long name — made a hard look at its acronym necessary. Referred to as ACS since Orton's time, the Society was sometimes confused with the American Chemical Society. On December 2, 1981, the Board approved ACerS (pronounced Ayesirs) as the sanctioned short form.

For 1986 the Society had revenues of $3.9 million and expenses of $3.7 million. But lagging membership recruitment remained a concern. During his term a few years earlier, President Robert J. Beals had estimated there was a potential for an additional 15,000 memberships from persons with established interests in ceramics.

ACerS expanded its continuing education program in 1987 with the acquisition of the California-based Ceramic Correspondence Institute, which offers practical education in the fundamentals of ceramics for technicians, production personnel, quality control personnel, salespersons, studio ceramists, managers and engineers of various disciplines.

At the time CCI offered 10 correspondence courses and five diploma programs, all targeted to individuals with little or no formal ceramic training. The courses were designed for home study completion.

The transaction increased the Society's ability to offer educational opportunities to technicians in the ceramics industry.

The long-standing contention that the Society emphasized pure science at the expense of the more "practical" aspects of ceramics was addressed in 1988 with the formation of a new class within the Society: the Ceramic Manufacturing Council. Its purpose was

> to promote the arts and sciences connected with ceramic manufacturing, including but not limited to engineering and technology, by means of meetings for the reading and discussion of ceramic manufacturing papers and for the publication of manufacturing papers, by coordinating multidivisional interest when appropriate.

Astronaut Bonnie Dunbar, a Society Fellow (pictured above with Society President Edwin Ruh at the 1986 Engineering Ceramics Division meeting in Cocoa Beach, Florida) took several items with her aboard the orbiter Columbia, *SST-50 in 1992, which now are displayed in the Ross C. Purdy Museum of Ceramics. Among them are the 1918 premier issue of the* Journal of the American Ceramic Society, *and a laser-engraved diamond wafer with the outline of the space shuttle and the names of the astronauts from the SST-50 mission.*

The next year the Board voted to increase the number of the Society's vice presidents from four to six. Because of the organization's growth, it was decided that vice presidents should be "functional" and focus on specific areas of the Society, such as research, applied science and engineering and manufacturing and business. These vice presidents would assure that the Society was addressing the needs of their particular communities.

After several years of hiring representation in Washington, in 1988 ACerS opened an American Ceramic Society office in the nation's capital. This bureau, manned by a staff member, was "dedicated solely to ceramic concerns and rapid, current information about legislative and regulatory action that could affect the ceramic industry."

"We need to be available in Washington to respond," Executive Director W. Paul Holbrook wrote to members. "Even more important, we need to be there to raise awareness, help shape the course of future actions, and encourage activities that may favor the Society, its members and the public that they serve."

Environmental regulations were some of those with the most potential to affect the ceramic industry. An environmental task force also formed in 1988

It was a busy spring in 1986 for Executive Director Paul Holbrook (fifth from right). Only a year after joining the Society in 1985, he oversaw the groundbreaking for a new headquarters building in the Brooksedge corporate office park and went to Paris with a group of Society travelers, including (seen here in the Latin Quarter from left) Ed Ruh, David Krim, Mal McLaren, Tom Root, Jack and Edith Wachtman, Barbara McLaren, Paul Holbrook, Beth Ruh, Dale Neisz, Terry Lindemer and Barbara Holbrook.

Alvin Weinberg, the 1997 Frontiers of Science & Society-Rustum Roy Lecturer, received a certificate from K.M. Nair (right), Lecture Committee chairman, and from Rustum Roy, for whom the lecture series was named. Weinburg has been widely recognized for his contributions to the development of nuclear energy, a field increasingly interesting to ceramists concerned with both the ceramics used in the nuclear industry and in environmental issues, including nuclear waste containment.

to gather and disseminate information regarding safety, health and environmental matters. The task force looked to members with regulatory expertise in areas such as hazardous chemical labeling, waste minimization, toxic chemical reporting and storm water discharges to volunteer their services.

Society members also had the opportunity to meet with regulatory agency representatives, members of the ACerS Government Liaison Committee, and the ACerS Washington representative during a 1990 symposium aimed at clarifying environmental regulations and compliance.

Guidelines for environmental responsibility were included in the 1990s proposed strategic directions. But although regulations created restrictions, they also created opportunities for new products. In 1972 President T.J. Planje predicted that, as manufacturers searched for more abundant substitutes for natural resources, "it would seem reasonable to anticipate that ceramic materials will have a greater role in our future technology and economy."

Almost 20 years later, in 1991, President Robert J. Eagan was encouraging members to "showcase the roles that ceramics play in solving environmental problems," such as eliminating pollution, detecting pollutants, treating effluents, capturing toxic materials and encapsulating waste.

Later that year, new President Dennis W. Readey enumerated still more environmentally positive uses for ceramic materials such as: high-performance batteries that would reduce electricity use, glass encapsulation of industrial and nuclear waste; and emission-reducing ceramic filters. These varied ceramic applications for energy conservation and waste disposal led Readey to conclude, "Protecting the environment is now accepted as an additional cost of manufacturing. What is becoming more apparent, is that preserving the environment may offer rewarding opportunities for new ceramic products and markets."

As the 1990s dawned and ACerS looked toward its centennial, information became a prized commodity. The Society, through its many publications and education programs, has always worked to keep members abreast of emerging new ceramic technologies as well as changes in existing fields. In 1993 the Society took its information services to the next level with the creation of the Ceramics Information Center (CIC), a resource for professionals with technical, bibliographical, commercial or general information requests.

Staffed by researchers and engineers who specialize in interpreting clients' questions, the CIC's goal is to provide the most thorough information possible, including company-specific product literature and consultants from the CIC's own "ceramics experts" database.

Yet another service-oriented initiative begun by the Society, the Ceramics Industrial Partnership Program (CIPP) promotes the competitiveness, industrial productivity and economic growth of U.S. ceramics producers. Sam Schneider, National Institute of Standards and Technology, led the effort to make the membership and board aware of the need for a program to influence the direction of research, development, and commercialization programs in ceramic-related areas; enhance the ACerS' service to industry; and provide before Congress and the Administration information and advocacy for policies and issues that affect the industrial climate of national and international ceramic companies.

CIPP was created, in part, to address the Society's growing number of corporate members. Tressler explained that when corporate members found themselves affected by regulations or presented with opportunities to work with the federal government, they often weren't sure how to proceed, how to protect themselves legally or how to reach their objectives. CIPP — with its information capabilities and advocacy orientation — was envisioned as the link between the two.

While CIC and CIPP were created to address the needs of present members, the MemberLink campaign challenged those same members to recruit new members. In the *Bulletin*, President Delbert E. Day praised the program for its efforts to "expand members' networking opportunities and [expose] new people to the wealth of ceramic-related information, products and services that the Society offers. This, in turn, enhances the overall strength and vitality of the Society."

Members employed a variety of creative recruitment techniques. Carlton H. Hogue, the 1995 cam-

Officers newly elected at the beginning of the Society's 100th year included (first row, from left) J. Richard Schorr, trustee-CMC; Stephen W. Freiman, president-elect; James W. McCauley, president; Paul F. Becher, treasurer; Warren W. Wolf, vice president at large; (second row, from left) John E. Marra, trustee-Nuclear & Environmental Technology Division; Thomas G. Reynolds, vice president-corporate and external affairs; Ronald E. Barks, trustee Engineering Ceramics Division; Dale A. Fronk, vice president-programs, meetings and expositions; Man F. Yan, vice president-publications; James E. Houseman, vice president-member services.

(Below) This planning focus group gathered in Orlando, Florida, to set goals for the Society in the last decade of the century.

By the mid-1990s, the phase diagrams so valued by ceramists were available in a convenient data base and were demonstrated at the annual meetings.

The floor of the exposition hall at any of The American Ceramic Society's annual meetings in the late 1990s illustrated a veritable United Nations of ceramics.

paign winner, included endorsed MemberLink applications in his Southwest Section newsletter. Said Hogue, "Actively recruiting members through MemberLink is my way of showing support for ACerS. . . . MemberLink expands on ACerS' strongest asset — its members. Everyone should MemberLink."

Initiatives such as CIC, CIPP and MemberLink helped balance the Society's long-range goals of retaining present members, keeping them informed in a rapidly changing industry and creating new service-oriented benefits to encourage membership growth.

At its Centennial, The American Ceramic Society carries into its second century some of the same issues it has faced since its earliest meetings — along with concerns unimagined by the founders but common to large organizations in the 1990s. But it carries as well the core values that have informed its membership and methods for 100 years: a respect for diversity of opinions, ideas and lively debate; a regard for fellowship; and a commitment to service to each other and the industry.

Finally, when everything else about the Society has been studied and reflected upon, one value holds it all together, unchanged for 100 years. As former President David Johnson, Jr. says, "The Society, for most of us, is where we find our friends." ▲

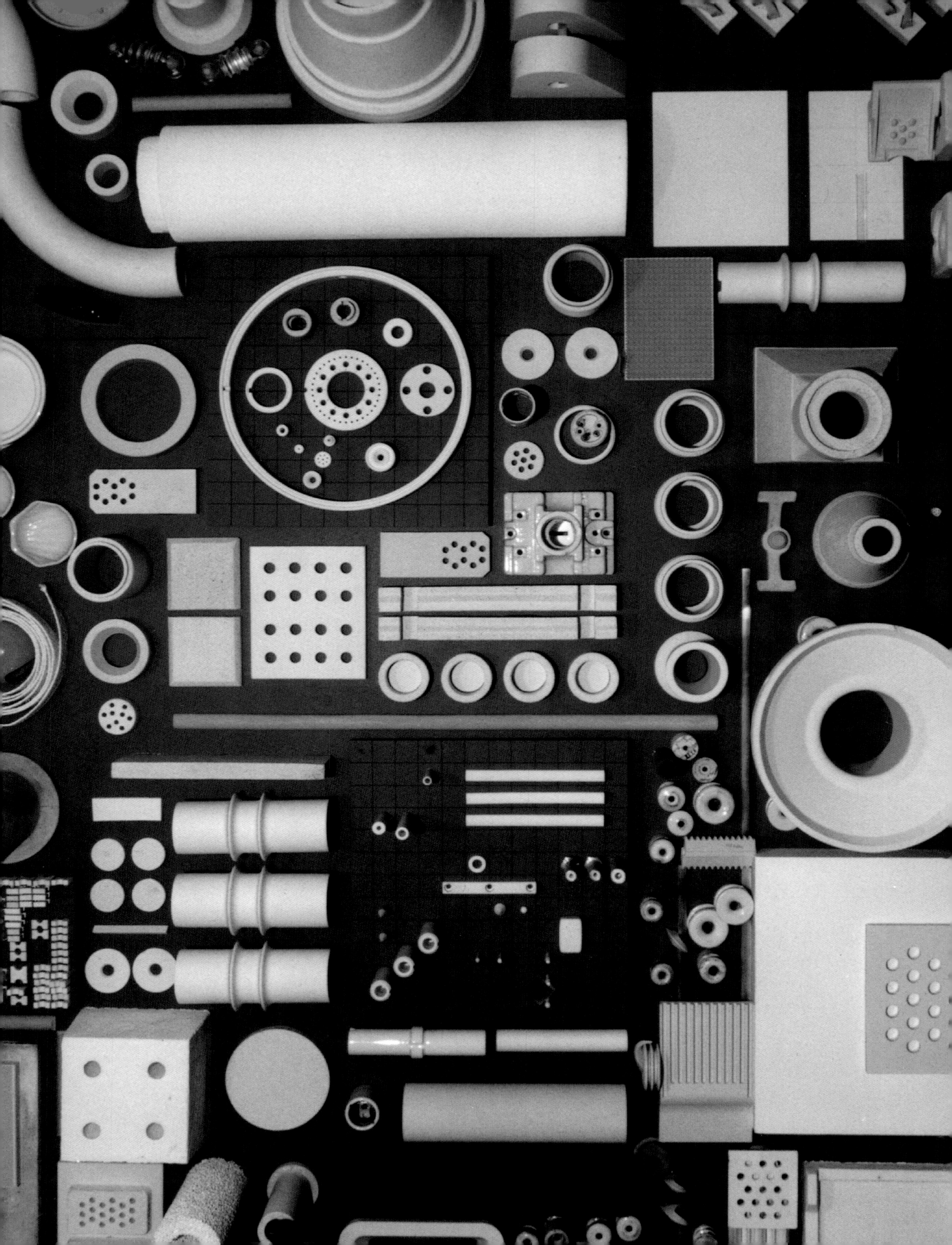

SOCIETY HEADQUARTERS

For nearly 20 years after the creation of The American Ceramic Society, its headquarters was the office of Edward Orton, Jr., in Orton Hall at The Ohio State University in Columbus. As the Society's first secretary, Orton conducted the Society's business from the campus until the United States became involved in World War I. • When Orton left to serve with the Army in 1917, the job of general secretary — and the Society's headquarters — passed to Charles F. Binns, dean at the New York State School of Clay-Working and Ceramics at Alfred University, New York. Like Orton, Binns operated out of his academic office. • By 1921 the Society had grown to the point that a volunteer secretary could no longer manage the Society's business. Ross Coffin Purdy was hired as general secretary, and beginning in 1922 free office space was provided in Lord Hall at The Ohio State University, where Purdy had been a member of the ceramics faculty. • By the end of the decade, though, the Society had outgrown the space in Lord Hall and moved into a rented office in the Gus Pallous Building at 2525 North High Street in Columbus.

(Above) The Society operated out of Edward Orton, Jr.'s briefcase during his tenure as general secretary, and the first Society headquarters were in space offered by The Ohio State University for Ross Purdy's use as an office.

(Left) The Edward Orton, Jr. Lobby of the Society's current headquarters in Westerville, Ohio, is home to two important pieces of ceramic art. It holds both the Star Chart *(opposite), a wall sculpture by Columbus, Ohio, artist Adrian Black, as well as the "Crystal Ball," seen at the far left of the photograph.*

The offices occupied half of the second floor on the south side of the building. In addition to office equipment, library and storage space, the headquarters contained an exhibit of ceramic objects representing clay, glass and porcelain-enameled ware from various times and sources. (The Gus Pallous Building was razed in the late 1950s.)

Shelves of contemporary periodicals (upper left) dominated the abstract office where staff maintained card files of abstracts indexed by title, author and journal reference. Like other rooms in the 2525 North High Street office, it was home to ceramic artifacts, as well, including (far right) a jar that was presented to Ross Purdy by the Czechoslovak Ceramic Society in recognition of his election to Honorary Membership.

By the early 1950s it was obvious that the Society needed a home of its own. In physically creating a national headquarters, the Society could establish a permanent center for the distribution of technical and scientific literature and firmly implant itself as the official organization of the ceramic world. At the same time, the headquarters would enhance the prestige inherent in Society membership.

Ground was broken for the built-from-scratch headquarters on April 2, 1954, at 4055 North High Street (later renamed and now remembered as 65 Ceramic Drive) on the northern boundary of Columbus' Park of Roses. Frank H. Riddle of Champion Spark Plug Co. in Detroit, a past president and chairman of the building fund committee, brought with him a ceremonial shovel made of ceramic spark plug material. This tool was used by President R.R. Danielson to dislodge the first spadeful of earth.

The 5,000-square-foot building in what came to be known as Ceramic Park — financed with a three-year, $100,000 fundraising drive — was dedicated in December of that year. At that time the Society had 3,800 members.

The building was constructed on three acres of rugged terrain. As completely as possible the structure was created from ceramic products — bricks, glass, terra cotta, porcelain enamel, tile, hollow wall tile, drain tile and sewer pipe. One interior wall was made of handmade brick at least 100 years old, retrieved from an old house that formerly occupied the site.

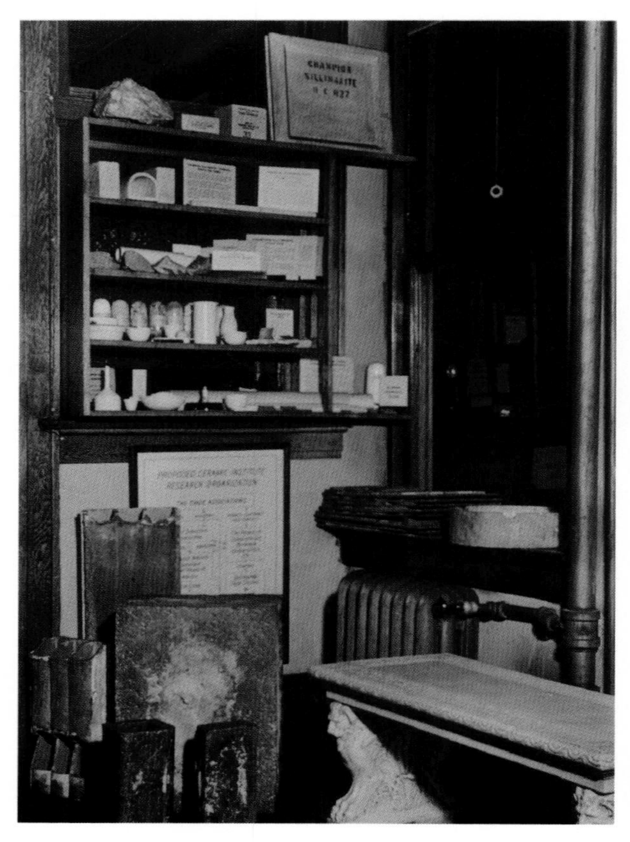

The Society's first official headquarters at 2525 North High Street in Columbus, Ohio, was occupied in 1926. It soon filled up with the Society's ceramic collections, representing production pieces mostly from American companies, as well objects d'art brought by the Purdys or Society members from their travels. Items shown here include a terra cotta bench made by Northwestern Terra Cotta Company; a block of Corhart electrocast mullite; flues from America's first Dressler tunnel kiln and, on the shelves and window ledge, an exhibit of Champion porcelain.

For all the pride Society members took in the structure, it did have a few unexpected idiosyncrasies. The liberal use of glass in the construction meant that

(Above) Guests at a 1946 reception, probably in honor of Ross Purdy's retirement, in the main office area of the Society's 2525 North High Street headquarters, mingled in front of the Society's collection.

(Above) The main office entrance at the 2525 North High Street headquarters held among its treasures color reproductions (seen above door) showing the history of optical glass, from the Bausch & Lomb company.

(Right) In 1930, the 2525 North High Street main office held a significantly smaller ceramic collection than it would in the years to come. Editorial and business functions of the Society were combined in Columbus in 1922, and the office staff worked hard to keep up with the growing volume of publications.

when the sun was out, the building became unbearably hot. Typically, when the outside thermometer hit 90 degrees, the staff was dismissed for the day.

Alice Alexander, who served as assistant to the general secretary — later renamed executive director — from 1960 to 1986, recalled that under those circumstances General Secretary Frank Reid, a man noted for his impeccable attire, was reduced to working with a wet towel draped around his neck.

Less than a decade later the Society was experiencing major growing pains. Membership had nearly doubled to 6,000 and the Society had an annual budget of $400,000. In 1963 a $290,000 expansion project was launched under the leadership of Jack Nordyke, general chairman of the building committee. By 1967 the effort had resulted in a major expansion that increased the High Street facility by an additional 12,000 square feet.

The building committee, seen here in May 1954 at one of its last meetings in the 2525 North High Street building, oversaw the construction of the 65 Ceramic Drive headquarters.

Among other benefits of this expansion was the creation of enough space to organize and install the Society's library, which immediately became a treasure trove of information for members.

Another bonus was the creation of a Board meeting room in the basement of the addition — the first time the Society had a room big enough for those official meetings.

By 1970 the Trustees, encouraged by rocketing real estate values in the Ceramic Park area, approved the construction of a new office building on vacant land in a valley along High Street, just below the Society's headquarters building.

It was believed by renting office space in this highly desirable neighborhood that the Society might create a source of steady income.

By the time the three-story, 35,000-square foot facility was completed in 1973 it already was 80 percent occupied. Among the tenants were several insurance agencies, a dialysis center, a physician, a business forms firm and an office of Western Electric.

No fundraising was necessary for this $800,000 project; it was underwritten by a commercial loan.

"The Ceramic Park building, particularly at this time, represents an excellent hedge against inflation and will provide for future space requirements. When the mortgage payments have been completed, the net

When ground was broken for the 65 Ceramic Drive headquarters on April 2, 1954, dignitaries from the Society used a ceramic spade, a tool that had been made of spark plug materials. Frank H. Riddle, a past president and chairman of the building fund committee, donated the spade, now in the Ross C. Purdy Museum of Ceramics.

After a fundraising campaign among members, especially corporate members, raised $100,000, a new headquarters building at 65 Ceramic Drive was built and dedicated in 1954. The installation of quarry tile on the building's front steps reflects the variety of ceramic materials used in the headquarters. As much as was structurally and functionally possible, the entire building was made of ceramic materials. Some of the companies that donated materials to the effort included: Wisconsin Porcelain, West Virginia Brick Co., Buffalo Pottery, National Tile & Manufacturing Co., Sheffield Brick & Tile Co., National Lead Co., Climax Fire Brick Co., Robertson Manufacturing Co., National Sales Corporation, Davidson Brick Co., Hydraulic Press Co., Industrial Colloids Co., Winburn Tile Mfg. Co., Wellsville Fire Brick Co., Aluminum Limited Sales Inc., Shenango China Inc., Elgin Standard Brick Mfg. Co., Gladding McBean & Co., Federal Seaboard Terra Cotta Corp., American Terra Cotta Co., Denver Terra Cotta Co., Northwestern Terra Cotta Co., Winkle Terra Cotta Inc. and Hammond Lead Products.

The Society's Board of Trustees posed for this group photo following their first meeting in the 65 Ceramic Drive headquarters building. Seen are (front row, from left): Charles S. Pearce, Herbert Insley, Paul V. Johnson, William O. Brandt, Robert Twells, Ray W. Pafford, R.R. Danielson, W.E. Cramer and Wurth Kriegel. (Back row, from left): Ralston Russell, Jr., G.H. McIntyre, Don Schreckengost, R.S. Bradley, Rolland R. Roup, J.S. Nordyke, John F. McMahon and W.E. Dougherty.

By 1965, the 65 Ceramic Drive headquarters needed an addition. At the groundbreaking ceremony (from left) Paul V. Johnson, John S. Nordyke, Howard P. Bonebrake, Elburt F. Osborn, Arthur J. Blume, General Secretary Frank P. Reid and George J. Bair used the Frank P. Riddle ceramic spade, used for the groundbreaking of the original 65 Ceramic Drive building.

Workers are laying foundation in this photo taken during construction of the 65 Ceramic Drive addition.

operating income will become income for the Society," Reid reported to the Board.

But by 1985 the Society was again bursting at its seams. Membership was just under 10,000 and the annual budget was more than $3 million. Board members briefly considered occupying the Ceramic Park Building to accommodate the Society's growth but were blocked by the long-term leases of the current tenants. In addition, a deep ravine separated the two buildings and prevented ready access from one structure to the other; to cross the ravine one had to walk around a long block.

The 65 Ceramic Drive headquarters was never replaced in the hearts of many members, even after growth required the Society to move again.

Instead they launched a search for a larger facility, and in the spring of 1986 signed a lease for a 22,000-square-foot facility being privately developed on Brooksedge Plaza Drive in Westerville, a Columbus suburb. It was located about six miles from the Ceramic Park buildings.

The new building was 28 percent larger than the old headquarters building and had 80 percent more useable floor space, providing room for up to 64 staff members, storage and an expanded Ross C. Purdy Museum of Ceramics. In addition, its location at the northeast corner of Columbus' I-270 outerbelt was easily reached from both highways and Port Columbus Airport.

A new building in a corporate office park development in Westerville, Ohio, became the headquarters both for the Society and the National Institute of Ceramic Engineers. When the Brooksedge building opened in December 1986, it had been built with a growing staff in mind, able to hold 64 staff members when the Society had only 45 at the time.

The move necessitated the sale of the 65 Ceramic Drive building and the Ceramic Park building — the loss of the Ceramic Drive building was keenly felt by many members who had worked and visited there and contributed funds and materials toward its construction. The proceeds from the sales of these properties were placed in a fund and used to partly offset lease payments on the Brooksedge address.

By 1990 the Society faced a tough decision. It could renew the lease on its headquarters in Brooksedge, purchase the facility, look for a different building or construct a new headquarters. By this time membership was up to 13,000 and the annual budget was $6.5 million.

Thanks to Executive Director Paul Holbrook and Director Gary Panek, who not only was ACerS director of administration but also a member of the municipal planning board in Westerville, it was discovered that

The lobby of the Brooksedge headquarters was the site for which the wall sculpture Star Chart *was originally commissioned.*

The columns and glowing colors of the main lobby and reception area at the Brooksedge headquarters reflected the trends in corporate space design in the 1980s, in contrast to the high, light space of the current headquarters building.

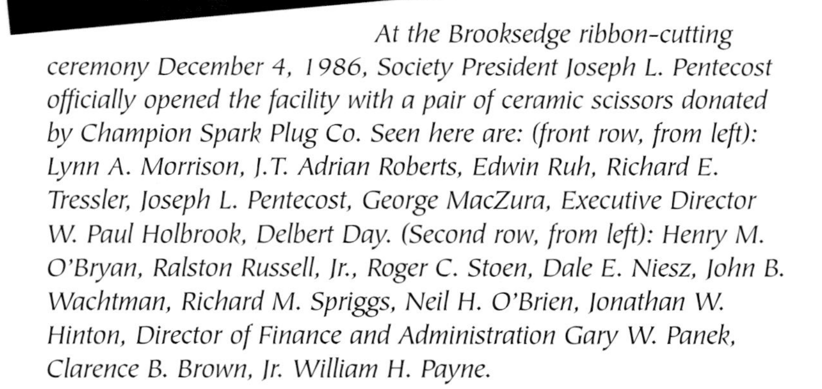

At the Brooksedge ribbon-cutting ceremony December 4, 1986, Society President Joseph L. Pentecost officially opened the facility with a pair of ceramic scissors donated by Champion Spark Plug Co. Seen here are: (front row, from left): Lynn A. Morrison, J.T. Adrian Roberts, Edwin Ruh, Richard E. Tressler, Joseph L. Pentecost, George MacZura, Executive Director W. Paul Holbrook, Delbert Day. (Second row, from left): Henry M. O'Bryan, Ralston Russell, Jr., Roger C. Stoen, Dale E. Niesz, John B. Wachtman, Richard M. Spriggs, Neil H. O'Brien, Jonathan W. Hinton, Director of Finance and Administration Gary W. Panek, Clarence B. Brown, Jr. William H. Payne.

The spacious Board of Trustees Conference Room in the current Society headquarters in Westerville, Ohio, was sponsored by Coors Porcelain Co. The brick wall sculpture depicting the history of ceramics was created by Allen Moran from materials donated by the Bowerston Shale Co. of Hanover, Ohio.

long-term, low-interest building revenue bonds were available for the construction of a new headquarters. Taking advantage of these bonds, the Society financed the construction of its current two-story, 35,526-square-foot headquarters on a 2.3-acre wooded site at 735 Ceramic Place, Westerville. The building, from which ACerS now faces its second century, was dedicated on December 7, 1991.

THE CENTER OF THINGS

In 1939 The American Ceramic Society analyzed its membership by geographic region, assigning a numerical ranking for each state and Canadian province based on the number of ACerS members residing there.

By this method, the center of population for ACerS members was 40 degrees 20 feet latitude and 83 degrees 41 feet longitude — a point approximately 30 miles northwest of Columbus, Ohio. ▲

The world's largest "Crystal Ball," weighing 700 pounds, sits in the current Society headquarters lobby. It was made and donated to the Society by artist Christopher Ries and manufacturer Schott Glass Technologies, Inc. The sphere is made of 35 percent lead glass, the same type used to create fiber optic cable.

The current Society Headquarters building in Westerville, Ohio, was officially opened in December 1991. It boasts 35,285 square feet, 55 percent more space than Brooksedge. Its construction was completely financed by $2.5 million in building bonds from the city of Westerville.

(Below) The James I. Mueller Memorial Library at the current headquarters building was sponsored by Society Past President William H. Payne and Katrina A. Payne. Open to the public, the library maintains more than 600 periodicals and thousands of books in several languages, with a professional library staff to assist members and other users interested in ceramics research and reference.

(Left) The Executive Director's office at the current headquarters building serves a dual purpose: It provides both a comfortable office space as well as a conference area.

MUSEUM CELEBRATES CERAMIC HERITAGE

Ross Coffin Purdy not only contributed greatly to the growth of the Society's membership and stature, but also to the wealth of beautiful things that have filled its offices for decades. The Ross C. Purdy Museum of Ceramics at Society headquarters now houses this extensive collection, made up of more than 700 ceramic pieces.

From 1921 to 1946, when Purdy served as the general secretary of the Society, ceramic wares from around the world that were given to The American Ceramic Society were documented in the Bulletin *during the 1920s and 1930s, and then put on display in the Society's offices at 2525 North High Street. Photos of this office show fine china and other ceramic pieces on display in long rows of shelves, although they sometimes shared space with office supplies. When Purdy retired in 1946, he left the entire collection to The American Ceramic Society.*

When the Society moved to its new headquarters building in 1954, most of the collection was packed away and placed in storage, since the new building had not been built with enough space to display the massive collection. In 1976, members began to ask about the collection again, and a committee was appointed to choose pieces and find a place for their display. This committee drafted the plan for the Ross C. Purdy Museum of Ceramics, which opened in 1981 at the headquarters building in a specially designed exhibit space funded principally by St. Gobain Corporation. Their gifts permitted the Society to purchase handsome rubbed-oak display cases for the items.

The Purdy gift created a nucleus for the Museum collection, but the focus of the Museum is also on significant ceramic objects from the past and present

The Ross C. Purdy Museum of Ceramics opened in 1981 after a five-year effort, initially proposed by then-president of the Society, Ralston Russell, Jr. to establish the funding and the collection. Seen from the room's doorway at the current headquarters, today's Museum is an astonishing array of ceramic objects from radomes rising like dragons' teeth (left), to the elegant dinnerware of American presidents (cases at right) to Dominick Labino's massive glass sculpture (circular object center). The public is welcome to the museum.

A terra cotta bust of Ross Coffin Purdy welcomed visitors to the Ross C. Purdy Museum of Ceramics in its Brooksedge location.

Display items in the Museum range from the very old to very new — from a ceramic brick believed to be from the Great Wall of China to the latest in biomedical ceramics used to replace human joints. The majority of the collection originated from Ross Purdy's 700-piece collection, which he gave to the Society upon his retirement in 1946. Among the handcrafted ware are examples from the Rookwood, Roseville and Weller Potteries, as well as from individual ceramists.

The Museum collection includes tile of many descriptions. Representative decorative and floor tiles from the collection include (clockwise from top): 1) Tile probably made to be used as floor pattern, made by International Tile Company, Brooklyn, New York, between 1882 and 1888; this tile is of the type made in England and shipped all over the world during the 1870s and 1880s. 2) Tile depicting male head wearing some type of armor helmet, made by Providential Tile Works of Trenton, New Jersey, between 1886 and 1913 when it closed. 3) A geometric floor tile made by the encaustic process by the United States Encaustic Tile Company of Indianapolis, Indiana, 1878-1882. 4 & 5) Two stove tiles made by American Encaustic Tiling Company of Zanesville, Ohio, circa 1890. 6) Tile made as a commemorative favor rather than designed to be installed as a decorative element, also made by American Encaustic Company. The back reads,"Interstate Mantel and Tile Dealers Association/8th Annual Convention/New York."

Neptune's Horses *is one of the Museum's 10 objects designed and produced by Frederick Carder (1863-1963), a Society Fellow who founded Steuben Glass Works, Corning, New York, and provided its artistic direction from 1903 to 1934. The colorless cast glass figure is an example of Carder's innovative work in cast glass made by the lost wax process.*

that represent achievements in technical and industrial ceramic production. Pieces in the collection cover a wide range of eras and ceramic styles. It includes pieces such as a ceramic brick believed to be from the Great Wall of China; an antique wall tile from Carthage, North Africa; up to modern industrial ceramic pieces such as one of the largest ceramic radomes on record and ceramic insulators. The bulk of the items were produced between 1860 and 1940. The Museum also holds exquisite ceramic art pieces, such as works by Frederick Carder (pioneer of the lost wax process of casting). His Neptune's Horses *is an example of glass-working at its finest.*

The collection includes works from a variety of artists and manufacturers, including: Roseville Pottery, Weller Pottery, Cowan Pottery, Hall China, Cambridge Glass, Homer Laughlin, Buffalo Pottery, Vulper Pottery, Catalina Pottery, Gladding McBean, Frankoma Potteries, Mary-Louise McLaughlin, the Overbeck sisters, Frederick Rhead, Muller-Luneville, Rookwood Pottery and Lenox China. ▲

(Above) The original storage and display area for most of Ross Purdy's collection was the main office space at the Society's first headquarters building at 2525 North High Street in Columbus.

Exhibits in Lord Hall at The Ohio State University showcased American pottery and other ceramic work and may have inspired Ross Purdy's desire to collect such objects. Here a 1925 exhibit of Rookwood Pottery contains work similar to pieces that are among the Museum's current holdings.

(Above) Museum visitors are surprised by the kinds of objects that can be made of ceramics. One display includes a tennis racket, hammer and protective heat shield, as well as the ceramic scissors used to cut the ribbon when the Brooksedge headquarters building was dedicated.

(Above) Discussing the more than 130 items on display at the 1981 dedication were (from left) Dorothy Wallace, long-time Society staff member; William H. Payne, Society president 1990-1991; Mary Gibb, long-time Society staff member; and President James I. Mueller.

(Left) Dignitaries present for the dedication included (from left) Past President William R. Prindle, President-elect Robert J. Beals, Past Treasurer Andrew Pereny, President James I. Mueller, and Treasurer Richard M. Spriggs.

(Below) Now in the Museum, the largest ceramic radome on record was presented to the Society by J.D. Walton, Jr. and N.E. Poules, Georgia Institute of Technology, who fabricated it.

(Below) Visitors to the 1981 dedication were greeted by staff members including Alice Alexander (left), assistant to Executive Director Friedberg.

During the 1929 Summer Excursion Meeting of The American Ceramic Society, the Baltimore-Washington Section hosted a gala dinner in honor of the visit of the Ceramic Society of Great Britain.

It may be something of a surprise to clay workers and scientists generally that an association composed of such men as form the roster of this Society should hold a regular formal meeting with absolutely no program in the way of papers or addresses. The experience of this Society has shown and abundantly proven that what makes the scientific or technical society chiefly valuable is not the papers and addresses and literature brought out in meetings . . . but the personal transfer of information and inspiration which accompanies these meetings. Most societies recognize this as a powerful factor, but few go boldly to the limit and say that this inspiration alone is sufficient. • The summer meeting is now pretty well established in this Society as the occasion for learning to know each other and by contact and by observation to acquire ideas from the plants of other ceramic and allied industries. The winter meeting is for the formal interchange of opinion by debate and papers on the technical problems of the day. Each meeting is of incalculable value and neither can be spared.

— *editorial appearing in* The Clay-Worker, *August 1900*

The Albert B. Sabin Cincinnati Convention Center in Cincinnati, is the site of the 100th Annual Meeting. Because thousands of delegates now attend the yearly event, the very close friendships among many members of earlier decades are shared by fewer people, but an overall spirit of warmth prevails. "I've heard us called 'The Kissing Society,' " recalled one Society veteran, "because we all get so darned close over the years. It's the one thing that really sets us apart from other technical societies, the family nature of our meetings."

THE KISSING SOCIETY

From the beginning, this opportunity to meet with other ceramists to exchange experiences and ideas was the heart of The American Ceramic Society. After 100 annual meetings, that hasn't changed.

> Our American Ceramic Society is truly a society by all definitions. It is an organization of people — warm, cordial people, who thoroughly enjoy both the professional and social camaraderie of our meetings. Many non-member attendees continually voice their surprise at this obvious difference between our meetings and those of other professional scientific or engineering organizations. May it always be so!

So wrote President James I. Mueller in 1982. And over the years Society members have proved him right again and again.

At least in part as the heritage of their colleagues in the National Brick Manufacturers Association, the early members of the Society liked to mix their business with pleasure. The most famed recreation in the early years of the Society was the acclaimed "Section Q" — basically a no-holds-barred bull session marked by much conviviality and, often, some real problem solving.

The early meetings of the Society always found time for recreation. On one memorable night when a business meeting was running late and threatening to impede on the theatrical presentation to which all the members had tickets, General Secretary Edward Orton, Jr. found it necessary to remind those present that the business of the Society took precedence over the theater.

The summer meetings during the early years generally allowed more time for social affairs, filled with fishing, golf and sports, balanced with plant tours.

The winter meetings deftly balanced work — the presentation of papers and the conducting of Society business — with entertainment. Wives who accompanied their ceramist husbands soon found themselves occupied with luncheons and organized shopping and sightseeing trips.

Eagerly anticipated highlights of early meetings were plant tours, which allowed members such sights as a Libby Glass Division employee blowing an electronic tube. A Society member from Sheffield, England, distinguished glass technologist W.E.S. Turner (center) and Mrs. Turner were on the tour.

The Summer Excursion Meetings began in 1899 and lasted until 1935. They emphasized fellowship and informal exchanges of ideas and experience. At the 1915 meeting, held near Niagara Falls, New York, delegates gathered for a photograph on the steps of New Castle Country Club. Edward Orton, Jr. (fifth from left, front row) and Ross C. Purdy (on step rail far left) were there.

Nationally syndicated newspaper columnist Ann Landers was the featured speaker in 1959, continuing a tradition of celebrity entertainment at the annual meetings.

By the late 1930s the annual meeting took on the trappings of an extravaganza. In addition to the lavish awards banquet, a highlight each year was a lively variety show attended by the delegates and their spouses. Over the years an impressive array of talent graced the Society's stage, among them the comedy team of Martin and Lewis, singer/dancer Ann-Margaret and, as a banquet speaker, Ann Landers. One year the entertainment was a "master pickpocket" reportedly capable of relieving men of their shirts and women of their brassieres without the victims even knowing they had been robbed.

Not everyone was delighted by this vaudevillian turn of events. "We have moved more and more away from a technical society and are now into the field of the luncheon and dinner clubs," groused one critic on the pages of the *Bulletin*. "True, the attendance has increased — almost doubled — but there is grave question that a better meeting has resulted."

High jinks and high spirits characterized meetings. At the Pittsburgh Section Meeting in 1950, toastmaster C.E. Bales, who had been Society president in 1943, demanded appropriate response for the Bleininger Award recipient.

As it turned out, the sort of talent conventioneers had been enjoying did not come cheap, and eventually the much-beloved variety shows became prohibitively expensive. They were replaced by dances in the mid-1950s.

The Annual Meeting of The American Ceramic Society has been held every year except one — the 47th Annual Meeting was canceled at government request because of heavy war-related demands on transportation and hotel facilities. Only the Board of Trustees, division and class officers and committee chairmen attended a much smaller gathering to participate in executive sessions.

Throughout the 1960s and 1970s, attendance at the Annual Meetings swelled. In 1960 Philadelphia witnessed what was then the world's largest gathering of ceramists, with a record 2,075 registered. Technical sessions had to be divided among three hotels, and sleeping accommodations were scattered through nine hotels.

The ratio of meeting attendance to the total membership was very high. At one point in the Society's history, 3,500 of 11,000 members showed up for a meeting. The same year only 10,000 of 120,000 members of the American Chemical Society registered for that group's convention.

As well organized as they were, the winter meetings could be exhausting and chaotic. A restful alternative

Formal portraits were an obligation of every annual meeting. At Atlantic City, New Jersey, in 1947, the officers and Board of Trustees included: (seated from left) E.H. Fritz; J.E. Hanson, retiring president; John D. Sullivan, newly elected president; John W. Whittemore, new vice president; W.E. Cramer, treasurer; and Charles S. Pearce, general secretary. (Second row, from left) A.K. Lyle, R.M. Campbell, L.C. Hewitt, F.R. Porter, J.H. Isenhour, E.M. Rupp, M.F. Beecher. (Third row, from left) J.R. Beam, C.M. Dodd, Theodore Lenchner, J.S. Gregorius, J.W. Hepplewhite.

was a fall meeting, held by many of the divisions and classes in various locations around the country.

One favored meeting spot was the Bedford Springs Hotel in Bedford, Pennsylvania. These gatherings were widely praised as among the best experiences the Society had to offer. Meetings occurred after the tourist season in the large old hotel, which was appointed with antiques, big metal beds, an 18-hole golf course, tennis courts and a trap shooting range.

"The meetings were never great technically — they were never intended to be," recalled Richard Eppler of the Bedford days. "You got to meet your colleagues in a more intimate way than at the annual meeting."

Russ Wood agreed: "Programming at the fall meeting was a joke. If you didn't have anything else to do, you went to the program. You would have four or five papers at the most." On the whole, delegates had other things to do, including golf outings, trap shooting and cross-country runs.

By the late 1980s it was clear that the annual meetings were just too big. The average gathering of this period featured nearly 40 concurrent technical sessions; more than 1,200 papers were presented; and the exposition area spread out over 15,000 square meters.

Only a few cities could accommodate the 7,500 registered delegates; even so, those attending were housed over a wide geographic area, making session hopping nearly impossible. It was a logistic nightmare for the Society's staff and volunteers.

To ease the logjam of the spring meeting, a second annual meeting — in the fall — was launched in 1989. The First Ceramic Science and Technology Congress was held in Anaheim, California, from October 31 to November 3, 1989, and was designed as an "embedded topical meeting," with most presented papers focused on prearranged topical subjects. Two more congresses were held, but attendance declined and they were abandoned.

Divisions were encouraged to hold viable, independent fall meetings, if they so chose, and participation in both fall meetings was possible if members had the wherewithal to attend.

In 1993 the Society held its first off-shore meeting, the PAC RIM Meeting in Honolulu. Held in conjunction with the First International Conference on Processing Materials for Properties, by the Minerals, Metals & Materials Society and the Mining and Materials Processing Institute of Japan, this gathering drew more than 750 attendees. The sponsoring organizations held independent yet joint meetings, and both independent and cooperative social functions. There were 708 papers read during the four-day event.

The Bedford Springs Hotel in Bedford, Pennsylvania, was a favored gathering site for division meetings for many years.

Louis J. Trostel, Sr. Society president in 1942, enjoyed a summer gathering at Bedford Springs in 1955. "When the Refractories Division went to Bedford Springs, you always scheduled a couple of papers so it wouldn't seem you just were going for the fun and games," recalled Robert J. Beals, Society president in 1982. "In contrast the Nuclear Division had papers all day long and then in the evening informal gab sessions where you had a chance to present a 15-minute talk on a state-of-the-art question. Nuclear was so serious."

A high point at many meetings was looking through the scrapbooks compiled by Josephine "Jo" Gitter, an indefatigable hostess and amateur photographer who delighted in capturing on film the social warmth of the gatherings, like this one at Bedford Springs in 1976.

The 1996 Society Reception at the Annual Meeting opened after the official ribbon cutting by (from left) W. Paul Holbrook, executive director; James W. McCauley, treasurer; Delbert E. Day, president; Carol Jantzen, president-elect; Rick Hughes and Douglas Bennett, representing the Indianapolis Convention & Visitors Association.

Delegates to the First Biennial Ceramic Manufacturers and Suppliers Workshop and Exposition in Louisville, Kentucky, in 1994 took time to socialize on a show boat.

President-elect David W. Johnson, Jr. took on a "Mr. Wizard" role for area television cameras visiting the Annual Meeting in 1994 in Indianapolis, Indiana, demonstrating the newly discovered superconductors.

THE EXPOSITION OF THE SOCIETY

The Society's first exposition was held during the 71st Annual Meeting in Washington's Sheraton Park Hotel on May 4-6, 1969, and attracted 116 exhibitors.

Advocates argued that the exposition provided an educational service to Society members, a marketing service to the raw material and equipment manufacturers and suppliers who supported the Society and were a source of substantial revenue for the Society.

Moreover, the argument went, the exposition narrowed the gap between the ceramic fine arts, science, engineering, factory management and production. But not everyone favored an annual exposition.

"We are facing resistance from some who feel an exposition held annually is prohibitive," General Secretary Frank Reid noted in 1970.

> Some favor an exposition only every second or third year. It is my hope to continue annually at a reasonably high level through the attraction of new exhibitors to offset the loss of companies choosing to exhibit on a biennial or triennial schedule. Unfortunately, a few companies are being extremely short-sighted by indicating that since they are not interested in exhibiting annually, the exposition should not be held every year. To point out the inconsistency of such theory, we are not asked to discontinue advertising in any given issue of the *Ceramic Bulletin* merely because a company does not choose to purchase space in that issue.

The 1973 Cincinnati exposition saw 10 booths being rented to British companies and groups, "an international endorsement of the value of the event," according to Society officers.

In fact, by the mid-1970s Reid could boast: "Revenue from the exposition enabled the Society to expand its services during a highly inflationary period and forestall an increase in member dues and subscriptions for a period of three years."

By 1986 the exposition was earning $300,000 a year for the Society. Also that year, the exposition was certified as an International Trade Fair and featured an International Business Booth to assist foreign visitors and exhibitors. It was staffed by interpreters fluent in German, French, Italian, Spanish and Japanese.

By the mid-1990s, the exposition featured about 250 international display booths representing traditional and high-tech ceramic products and process.

The Centennial exposition will include 180 exhibitors, 15 of those international participants. In addition, an "Internet café" will offer computer access to participants.

SECTION Q

"Ask any member of the first five years what he enjoyed most [about early annual meetings] and he will say 'Section Q'," Norah W. Binns wrote in 1923.

"At the outset, when there was a warm, cordial intimacy among all the members," reminisced a member in a 1934 article in the Bulletin. "We would hold uncatalogued sessions in some quiet rathskeller and, with corncob pipes and a stein of beer, would hold informal, enthusiastic meetings where intimate experiences could be safely exchanged that our employers would not permit in open sessions. Stanley Burt, the able technician of the Rookwood Pottery, the pride of American ceramics, was a leading spirit in the camaraderie of Section Q, where the formality of Professor Orton was notably absent."

Norah Binns explained Section Q this way:

> The need for recreation was not satisfied by dancing and moving pictures. It was talk and more talk that these men craved. Formal papers were given morning and afternoon, but in the evening, in a private room or around a table, refreshments of the good old variety were made available, cigars and pipes were lit, and Section Q was on.

Ellis Lovejoy, a Charter Member of the Society and its president in 1913, was an early colleague of Edward Orton, Jr. and served the fledgling Edward Orton, Jr. Ceramic Foundation as its director of research from 1928 to 1933. By all accounts, his jolly humor contributed to setting an upbeat tone for Society gatherings from the beginning.

Many of the members were working alone so far as scientific sympathy or understanding was concerned. All of them had problems, all of them had a little knowledge. Put together, the bits of knowledge gradually began to penetrate the problems. Section Q was the earliest form of ceramic cooperative research.

Binns gives this anecdote that suggests the conviviality that dominated these sessions:

> On one occasion a charter member had to make a speech at the banquet of another organization meeting in the same city. He stayed at his beloved Section Q as long as he could, then rushed away to keep his engagement. A loyal friend proposed that their group should support the speaker by going in a body to hear him, and so they did.
>
> But before their companion's speech they had to listen to an interminably long address by someone else. Worn out but at last about to get away they looked for the man who had suggested coming to the banquet, only to find he had retired long since and was soundly sleeping.
>
> To make the punishment fit the crime, an urgent call for five a.m. was left at the clerk's desk for the sleepy one.

Although the Section Q meetings were history by 1946 when a group of members were snapped at a reception at Society headquarters, conviviality and conversation — and attention to proper attire — were still highly prized.

Ironically, it was the rapid growth of the Society that rang a death knell for Section Q. One Society historian observed:

> As the members grew in numbers, coming from distant places and without many personal acquaintances, and especially with the growing college groups with their very intimate and close personal ties, and because the trading of shop experiences was more largely by correspondence throughout the year, Section Q became altogether social and generally in groups without any other purpose than to renew personal acquaintance and to again talk over past times. The new members . . . found little of interest in these latter-day Section Q meetings.

By 1910 Section Q had passed into Society legend. ▲

SECTION Q ENCOURAGED BOTH FACT AND FABLE

In an issue of* The Clay-Worker, *J.E. Randall reported a story in the true spirit of Section Q.

. . . The party then boarded a car for Lincoln Park and . . . a refreshing dinner on a cool piazza in the Park. It was here that Ellis Lovejoy got in his good word and told his remarkable fish story, which he claims is absolutely true. It seems that he and Professor Orton had been out shooting bears all morning in the mountains of Colorado, and, becoming tired of their merciless onslaught, turned their attention to fishing. They were not very successful and were about to return to camp disgusted with their luck, when Lovejoy hit on the scheme of putting whisky on the bait. This worked successfully and they were kept busy pulling in the fish, when Lovejoy accidentally knocked the bottle overboard, and that settled it; the fish came up out of the water of their own accord and climbed into the boat. It may be hard for the readers of The Clay-Worker *to believe this, but Mr. Lovejoy says he will stake his reputation as a brickmaker on the story. After lunch Mr. Warder took the party through the park in the park vehicle, after which they returned to the Hotel Kaiserhof, which was headquarters during the meeting. During the evening there were several Section Q meetings, the most important of which was held at Bismarck Garden.*

Another Version of the Orton-Lovejoy Fish Story

My Dear Randall:

I can readily believe that your correspondent held a Section "Q" meeting at Bismark's [sic] Garden and that he dreamed that fish story. I am informed that it is a place where dreams are made. The story is preposterous, and when you have the facts before you I think you will agree with me.

Orton and several others spent three weeks in Northern Colorado climbing mountains, and it was my privilege to be one of the "others." Our first camp was under Long's Peak, well up to the timber line and at the foot of a snowbank, far above civilization and fish. We were there ten days, and not a single fishing rod was taken out of its case. Put yourself in my place.

How much of that "bait" do you think would be left at the end of that time? As a matter of fact, all our bait was used to rub our aching joints at the end of the first day's climb and we had to fall back on bacon rinds at the end of the second and subsequent climbs.

Your correspondent dreamed that boat. Our second camp was at Bear Lake, at the foot of Hallet mountain. There is hardly a trail into the lake, and so difficult is the route that we had to carry the packs and pack animals as well over part of the way. None of us had any boats around loose in our pockets on that journey. Whatever route you take you select a different one coming out and wished you hadn't.

It is my painful duty to tell you the truth about those fish.

When we reached Bear Lake our sun-kissed faces (I like that expression, and hope it is original. After the third kiss Orton's face had a maidenly blush which would have made a whole palette of ox-blood reds look like neutral tints.) had reached the second peeling time, and thinking of Chicago and home, we were inclined to sit on the rocks which border the lake in the shadow of the mountains and tone down our complexions. For occupation we fished and smoked, and at intervals removed, gently of course, sections of wornout cuticle to aid in the toning-down process.

The fish scorned flies made of fuzz, feathers, and hair, nor would they look at worms, however artistically we baited them. But we looked so comfortable and happy up there on the rocks that the fish came up from their homes in the dark, cold depths to share our pleasant positions, climbing the lines and sliding down the poles to reach us. They came in such numbers and were such leviathans that we had to retire to the wood to avoid being crowded off the rocks into the water. Of course, we took a few of the smaller fish — such as we could carry — with us for company, and they are still with us. There is all there is to that fish story. . .

. . . Call off your correspondent and to the next summer meeting send a Munchausen or, better still, come yourself.

Yours truly,

Ellis Lovejoy

P.S. - I have purposely drawn out this tale of facts that you may not be tempted to print it. Cold printers' ink gives me the "creeps." ▲

(Right) A gag mug with monks drinking as decoration, stamped on the base "Mid-Summer Meeting, American Ceramic Society, Canton, Ohio, July 25-27, 1921" was provided to delegates, compliments of the Sebring Pottery Company, Sebring, Ohio.

DIVISIONS ENCOURAGE GROWTH

The formation of industrial divisions, which was initiated in 1918, has been called the single most significant step in the Society's history.

In the Society's early years, the membership was small enough — and ceramic technology of a limited enough scope — that the needs of just about all members could be met with a single meeting.

But as the ceramic field grew more complex and varied, it became apparent that the organization would benefit from being divided into specialties.

Initially, at least, the divisions were primarily focused on products — refractories, whitewares, heavy clay products and so on.

The first divisions to organize were Enamel, Glass, Terra Cotta and Refractories, all in 1919. In 1920, the Board approved formation of the Ceramic Decorative Processes Division, while Heavy Clay Products and Whitewares held their first meetings in 1921. The Materials and Equipment Division held its initial meeting in 1931, the same year the Heavy Clay Products Division approved a plan to merge with the National Brick Manufacturers' Association to form the Structural Clay Products Division.

Over the last 80 years new divisions have been created, some have faded away, many have changed names or been folded into another. At various times in the Society's first century, most divisions have gone through periods of stagnation and dwindling membership. Often these situations have been answered by transforming an existing division or creating one, more or less from whole cloth, to represent new branches of the diverse field of ceramics.

Originally, divisions were set up on the basis of product, with each division acting almost as a separate society with its own officers, programs and meetings. Following this reorganization the Society grew rapidly, the programs expanded and the major leadership efforts turned toward the publications.

Divisions greatly stimulated the Society's involvement and support of these fields, attracting able scientists and engineers who might otherwise not have joined the Society, and thus serving the members better by making sure that the best people in these fields would attend meetings and share information of vital interest to other ceramists.

While the divisions fueled much of the drive of the Society, they also created a certain amount of, well,

Cove Point, Maryland, in 1929 was the site of the first Glass Division Summer Meeting (later called Fall Meetings: "Proximity of the date to the autumnal equinox, and the varieties of weather encountered, make nomenclature difficult" wrote one chronicler). Francis Flint, Division head, invited a select group to his summer cottage, including: (standing, from left) G.W. Cooper, Dale Schurtz, Herbert Insley, A.K. Lyle, William C. Taylor, Wilbur F. Brown, Arthur Q. Tool, G.E.F. Lundell; (seated from left) E. Ward Tillotson, Frank W. Preston, Jesse T. Littleton, Francis C. Flint, C.G. Peters (as photographed by George E. Merritt).

By the late 1930s, the Glass Division meeting had grown considerably, to judge by one photographed at a hotel in upstate New York near Bath. Ross Purdy is fifth from the right in the front row.

Seminars and short courses have provided countless ceramists with new or expanded knowledge in their own and related fields of interest. Among many outstanding educators in the Society, one exemplar was Gilbert C. Robinson, whose broad career centered on the brick industry but encompassed many areas of ceramics. He provided hundreds of lectures and seminars for ceramic professionals and his presentations were the staples of many Society meetings.

division. It has often been said that the divisions sometimes became too consumed with their own activities and weakened the common bond that held together the members in the early years of the Society.

For most of the Society's first century — until the early 1980s — members could join only one division. But after years of debate it became increasingly clear that single-division membership was short sighted. Modern ceramics, after all, is not a series of neat little compartments, each based on a product and neatly insulated from the next, but rather a vast scientific community of interrelated interests. With the growth of so-called materials science, this truth has become inescapable.

It was recognized that the information developed in one division could be of tremendous interest and importance to members of the other divisions. It was also acknowledged that multidivisional memberships would provide a cross-fertilization of ideas among divisions and would nurture interdivisional interests. Accordingly, the Society's rules were amended to permit members to enroll in up to three divisions (though a member could hold office in only one division at a time).

For those who desire rapid change, though, the Society could prove frustrating. Traditionally it has been cautious about creating new divisions to address new ceramic technologies. In part this may be because Society members already find themselves pulled in many directions. In addition to his or her divisional memberships, a ceramist may also belong to a local section, and quite possibly to the National Institute of Ceramic Engineers, the Ceramic Manufacturer's Council and the Ceramic Education Council, all classes within the Society.

"The formation of extra divisions should be avoided," observed Harry E. Davis, chairman of the Society's Committee on Divisions and Classes in 1956, when a move was afoot to create an Electronics Division. "If we were to break up into groups based on the end-point use of products, we would need a complete reshuffling of our present divisions."

Added then-General Secretary Charles S. Pearce:

> I hope you recognize the difficulties of operating separate divisions. At the present time the Society has eight divisions and every additional activity of this kind compounds the trouble at meetings and does not seem to contribute much to the effectiveness of a program of interest to the different individuals.

Poster Sessions have become increasingly important over the years as the number of papers available to be shared has grown exponentially. Poster presentations are especially suitable to research that can be explained in figures or photographs without requiring an accompanying lecture. The Poster Sessions have achieved sufficient import that, in addition to being open to members every day of the annual meeting, posters command a half-day during which no other presentations are scheduled, making it possible for everyone to view the posters.

Nevertheless, new divisions were created in the ensuing years. And it seems likely that with the development of new ceramic substances and with new applications for existing ceramic materials, other divisions will continue to be created from time to time.

In 1998, the Society's current divisions, in alphabetical order:

BASIC SCIENCE DIVISION

The Basic Science Division has the most diverse membership of any division in terms of fields of training, variety of research and development activities, varied subject matter and materials, and its mix of organizational affiliations, which span universities, government, nonprofit groups and industry.

Created in 1951, Basic Science is unique in that it is not product- or process-oriented. Instead, Basic Science appeals to members interested in the fundamental facts, principles and theories that underlie the synthesis, processing, characterization and properties of ceramic materials. Oriented heavily toward scholarship and research, the division is the Society's prime generator of technical papers.

Initially it was suggested that the new division be designated Physical Chemistry, but that was viewed as too narrow in scope.

A 1978 survey of division members revealed that nearly 30 percent were trained in ceramics, about 20 percent were trained in chemistry, and about 10 percent held degrees in metallurgy. Fully 60 percent of division members were engaged in either research or teaching. Basic Science members historically have contributed nearly half of the content of the *Journal*.

From its inception, the Basic Science Division has been unique, since it cut across the usual division structures to draw members from throughout the Society. In fact, at the time of its creation, before multiple division memberships were allowed, some members urged the Society's trustees to permit new Basic Science members to retain their memberships in their original divisions, lest a mass defection to Basic Science result.

Largely because of this concern, in 1971 it was suggested that the division be converted into a class of the Society. Supporters of the idea noted that a wider membership would be attracted to a Basic Science Class, and that a class member could also belong to a division. The drawbacks would be that as a class, Basic Science would lose its roles in the review system and program organization. The conversion to a class was

The Basic Science Division, shown meeting in 1974, has the most diverse membership of any division in the interests and training of its members. But by the early 1980s, many people were chafing under a "one division only" membership rule. Changes in the industry almost demanded multidivision interests, so the rules were changed to permit membership in as many as three divisions at once.

Even at Whitewares meetings, it's not just the plate but what there is to put on it that draws the crowd. From the inaugural meeting of the Society, good company and nice "refreshments" have been important to Society meetings.

At the 1976 Refractories Division meeting, scrapbooks of previous years' activities absorbed delegates and companions during the social hours.

Even though the nation's roadways and cities depended from the beginning of the century on ceramic materials, the Cements Division first proposed in 1918 did not become a reality until 1971.

rejected. But largely because of the existence of the Basic Science Division, the Society approved — after several rejections — dual divisional memberships in 1980.

CEMENTS DIVISION

Though first proposed in 1918 and then again in 1953, the Cements Division was not created until 1971 when it became obvious to all that it could serve as the national forum for interaction between persons interested in the science and technology of cements, a multibillion dollar industry in the United States alone.

Prior to that time the American Concrete Institute, the American Society of Civil Engineers and the Highway Research Board had dominated the field, but not with the same technical and scientific emphasis that could be provided by the Society.

At the same time it was obvious that inorganic cementitious materials (including portland and aluminous cements, among others) are ceramic materials, and would benefit from knowledge related to the manufacture and use of other ceramic materials.

Significant among the Division's projects is the publication of *Cements Research Progress,* an annual survey of cements science literature. An invaluable resource to professionals in the field of cements research, the series presents full bibliographic citations from the year's published works, as well as precise summaries and commentaries concerning the works written by dedicated Division members.

DESIGN DIVISION

The Ceramic Decorative Processes Division — often referred to as the "Art" Division — was created in 1922, a time when a good deal of ceramic activity actually revolved around "fine arts" ceramics.

Orton and the other Society founders clearly saw their organization as one involved in both the arts and science; their awareness of the importance of art to the ceramic world was reflected in one of the Society's first votes, which was to bestow an honorary membership on Mrs. Bellamy Storer (formerly Miss Maria Longworth), the founder and guiding light of the world-famous Rookwood Pottery. Yet for most of its

A juried competition created to showcase examples of ceramic art, the Centennial ceramic art exhibit encouraged entries from around the world. Selected pieces included Can't Get a Handle on It *by Richard Montgomery.*

history the Art — later Design — Division has had an uneasy relationship with the rest of the Society. Comprised in large part of fine arts ceramists, hobby potters and fine arts educators, the division has for decades been an uncomfortable fit with the scientific and industrial types who dominate the Society's membership.

In the early years the division organized art exhibits in connection with the annual meetings. But there were many years of struggle, due to low membership and high turnover. The division's small size made mounting presentations and meetings difficult. At various times in the last 80 years the division has undergone several reorganizations intended to give it a boost.

Still, the division's fundamental right-brain, left-brain conflict with the greater Society remains to a degree. "Many ceramic artists and craftsmen no longer feel a kinship to the largely technical society we have become," Design Division member and Society President Stephen D. Stoddard wrote in the *Bulletin* in 1977, at a time when the division had held no sessions at the Annual Meeting for nearly a decade.

In the late 1970s headquarters became aware of a group of about 150 studio potters, teachers, students, commercial potters, fine arts ceramists and designers concentrated on the West Coast. Calling itself the Design Division of the Southern California Section of The American Ceramic Society, since 1948 it had operated at the periphery of the society, without official recognition. After a flurry of discussion, it was decided that the group would be called the Design Chapter of the Southern California Section of the Society.

Though it focused chiefly on art pottery and, to some extent, dinnerware, Design gradually expanded its vision to include industrial design in the belief that good design is sometimes of greater importance to sales success than technical quality.

The Society's recent purchase of *Ceramics Monthly* magazine and the introduction of two new publications, *Pottery Making Illustrated* and *Potters Guide*, mark a renewed effort to serve this vital part of the ceramic community. The Society will also cosponsor the 1999 National Council of Educators in the Ceramic Arts (NCECA) meeting, along with The Ohio State University.

GLOSSARY for SPEAKERS

Slides — pictures, diagrams, and charts prepared by the author and literally thrown on the screen. Note: Some speakers are so chummy with these that they are apt to talk to them instead of to the audience. Rather startling effects can be had with them by (a) throwing them on and off the screen at a rapid rate, or (b) by placing freshly made slides in the projector (pyrotechnics).

With tongue-in-cheek observations and drawings, Henry Plod-Kin, a.k.a. J.F. McMahon and L.R. Bickford, brightened up the 55th Annual Meeting with their "Glossary of Terms for Persons Giving Technical Papers." So timeless were the pair's observations that many of them appeared again in the program for the 73rd Annual Meeting.

Putting contests, like this one from the 1994 Annual Meeting, have been a perennial favorite. Held in the hotel lobby at some annual meetings, with ceramic putters as prizes, the contests have drawn not only members but occasional passers-by.

Luncheons, like this one in 1976 at Bedford Springs, and bridge parties have been favorite entertainments for decades, with prized favors donated by ceramic companies.

A combined Materials and Equipment Division and Whitewares Division meeting at Bedford Springs in 1976 hosted an elegant dinner.

Some members complained that dinners had taken on a "vaudevillian quality" with the addition of celebrity entertainment, but dancing remained popular.

COMPANION PROGRAMS

For decades, the "ladies" (as the wives of members were almost always described in Society publications and lore) may have had the best of the meetings — days spent renewing friendships at luncheons, teas, bridge parties and fashion shows, with no tedious papers or contentious administrative sessions to interrupt the fun. Evenings went to dinners and parties; at Annual Meetings there were the gala dinners, such as the 1947 extravaganza in Atlantic City, described in the Ceramic Forum *newspaper by Edward Marbaker, later editor of Society publications:*

> *During the late evening the whole crowd of ceramists — and it was a crowd — assembled in the Convention Hall ballroom for what was called Boardwalk Night. The entertainment consisted of a very fine performance by an accordion virtuoso (didn't get his name but it was the first time this writer every enjoyed that kind of music), an excellent dancing team, impersonations by the master of ceremonies, and the gyrations of a whole family of roller skaters. The show was put on by Howard Lanin and it was voted a great success by everybody present. The evening wound up with a drawing and the presentation of a large number of prizes contributed by ceramic firms from all over the country. These prizes went to the women, but one lucky man, whose card was the thirteenth drawn, won the Schweitzer door prize of $100.*

Over the years times changed, and the Society changed to meet the membership's evolving expectations. Variety shows were replaced with simpler dinner programs. "Ladies Entertainment" became "Companion Programs" and accommodated the fact that a larger number of spouses and companions every year were men at the meeting with women who were Society members.

The changes may have meant fewer fashion shows — but time has not diminished the pleasure that members and their companions at the meetings take in seeing old friends year after year. ▲

ELECTRONICS DIVISION

The Electronics Division was created in 1957, largely of persons who previously had belonged to the Whitewares Division but who believed that electronic ceramics was fast becoming a discipline unique unto itself.

Among those petitioning for a new division were people interested in ferromagnetics, ferroelectrics, isolators, whitewares for electrical use, glass-to-metal sealing and the electrical properties of glass.

At the same time, manufacturers of traditional whiteware observed that their divisional meetings were being inundated with technical papers on electronics, which were of little or no interest to them. When quizzed by Society headquarters, most of these traditionalists said they would be pleased to see an Electronics Division set up its own shop.

When Sputnik *went up in 1957, the nation demanded more science to keep up with and surpass other nations. The field of electronics burgeoned in the years following and the Electronics Division, formed in 1957, was ready for growth.*

ENGINEERING CERAMIC DIVISION

The Enamel Division was one of the Society's original divisions, and for its first 40 years was exclusively concerned with glass and ceramic coatings on metals which included cast iron, steel, aluminum, stainless steel and other nonferrous and refractory metals.

By the late 1950s and early 1960s, though, it became obvious that changes were afoot. Relatively little advancement had been made in the traditional area of enameling, and the division was experiencing a slowdown in of new memberships; meanwhile numerous new related technologies were developing.

Individuals such as R.S. Sheldon, then secretary of the Enamel Division, recognized that The American Ceramic Society was quickly evolving from an industry-based to a technology-and science-based membership. "A new or revised scope is a must for our division," Sheldon wrote, "and a change in name should be accomplished as quickly as the mills can grind through the maze of red tape."

The new name — official in 1960 — was the Ceramic-Metal Systems Division.

Among other things, plastic was beginning to replace porcelain enamel in many large consumer appliances. Thus the purview of the division was expanded to include glass-to-metal seals, ceramic-to-metal seals, high-temperature refractory coatings on alloys, special oxidation-resistant coatings on refractory materials, cermets and other metal-ceramic composites.

An Electronics Division factory tour in 1959 to see spark plugs manufactured was not like the dusty excursions of the mine and plant tours of an earlier, clay-based ceramics era.

GLOSSARY for SPEAKERS

Heckler — the guy who sits in the audience waiting for you to say something so that he can dispute you. He can be readily recognized by (a) the wrinkled brow, (b) an unsmiling countenance, and (c) diligent note taking. Note: speakers find difficulty in differentiating between pseudo-hecklers and the actual specie.

Large or small, meetings attracted a similar cast of characters — some more welcome than others.

The Glass Division has always been one of the Society's largest divisions, and it is also the single largest and most important glass organization in the United States. This 1960 meeting drew a considerable crowd.

The renamed division bridged numerous gaps between ceramic activities and the metal industries, and opened up for the division new ceramic materials for aerospace technology, architecture and building construction, wear-resistant materials and electrical phenomena in ceramics.

By 1985, the landscape had once again changed enough to warrant a divisional name change — it became the Engineering Ceramics Division.

Frank D. Gac, the first chairman of the Engineering Ceramics Division, maintained that the shift in emphasis allowed the division to meet the challenges of advanced material concepts and new engineering systems, including ceramics for energy conversion systems, friction and wear, brittle material design, materials for advanced space transportation, ceramic cutting tools, grinding and abrasives and advanced processing technology.

In a dramatic departure, the Ceramic-Metal Systems Division, led by Jim Mueller, Jim McCauley, Jerry Persh, John Buckley and others, initiated the "Cocoa Beach Conference" on Composite and Advanced Ceramic Materials in January 1977. Approximately 89 people attended that first meeting, and attendance grew to more than 1,000 by the late 1980s. It was the obvious success of the initial meeting that eventually led to the name change from Ceramic-Metal Systems Division to the Engineering Ceramics Division.

GLASS AND OPTICAL MATERIALS DIVISION

The Society's Glass Division is the single largest and most important glass organization in the United States.

One of the original divisions created in 1919, the Glass Division was also responsible for several Society "firsts." It is believed that the division initiated the idea of meeting as a group independent of the Annual Meeting in 1929 when Francis Flint, then division chairman, and about 10 members met at Flint's cottage on Chesapeake Bay for a weekend of golf, tennis, hiking and other forms of camaraderie. The idea was soon taken up by other divisions, notably Refractories and Whitewares.

The Glass Division initiated fall technical meetings in 1930, and these became a regular part of the calendars of all divisions.

In many ways the Glass Division became a microcosm of the Society as a whole. For years before World War I the division was led by a group dubbed by General Secretary Ross Purdy as "the Moguls of Glass." Under their leadership, the *Bulletin* reported in 1946, "the Division has become much like a fraternity, in which a warm cordiality tempers the coldness of technical discussion, and firm friendships grow and first names are used."

And the Glass Division early developed international connections with strong participation in the International Commission on Glass (ICG), a worldwide federation of glass technical societies. The Society was a charter member of the ICG in 1933, along with glass societies from Germany, Great Britain, France, Spain and Italy. Today there are about 30 member societies.

The late 1980s was a curious period of relative stagnancy in the division — at a time of rapid expansion of the Society as a whole and of glass research in particular. Division leaders rallied to stop the loss of members interested in the new, emerging area of glass — including fiber optics, sol-gel, fluoride glasses and other nontraditional, noncrystalline materials — to other divisions and other societies.

Today the division, in 1990 renamed the Glass and Optical Materials Division, deals routinely with products and techniques undreamt of just a generation ago. Among its concerns are glasses and glass-ceramics important in electronics packaging, glass fiber amplifiers, glasses with nonlinear optical properties (heavy-metal oxide and semiconductor-doped glasses), glass coatings, hydrogenated glassy silicon in active matrix liquid-crystal displays, glass as a photoconductor and in solar energy applications.

In 1932, the Glass Division's leadership, dubbed the "Moguls of Glass" by Ross Purdy (seated right in front of column, side view), met at the Woodmont Rod and Gun Club near Hancock, Maryland, where they "enjoyed the facilities of the club . . . and inspected the duck ponds, the turkey hatcheries, the pheasants, and the extensive parks for deer. . . . The program consisted mostly of informal conferences," recalled S.R. Scholes in his 25-year history of the division. Francis Flint is fourth to the left of Purdy, second row.

MATERIALS AND EQUIPMENT DIVISION

The Materials and Equipment Division, the first division added to the original slate put in place in the late teens and early '20s, was organized and dedicated during the 33rd Annual Meeting in 1931.

In his presidential speech that year, retiring President Edward Orton, Jr. spoke of the formation of this division as one of the concrete steps that would increase the power and effectiveness of the organization. He pointed out that:

> Some of you perhaps have not yet sensed fully the need of such a division, reasoning that all the divisions have in the past given place to papers and discussions on such topics, and that under that procedure new materials and new equipment have been brought forcefully to the attention of those most directly concerned with its use. On the other hand, ceramic materials and processes in the nature of the

Executive Director Paul Holbrook and longtime Society member Carroll Kay, Kyanite Mining Corp., called a bingo game in 1994 at the first Biennial Ceramic Manufacturers and Suppliers Workshop and Exposition. The workshops, similar to annual meetings, focus on the interests and needs of Society members and allied organizations concerned with ceramic materials, processing and products. These workshops have been important enough to members that the applications divisions have suspended their fall meetings in the years when workshops are held.

The head table at a Materials and Equipment Division joint meeting with the Whitewares Division at Bedford Springs in 1962 looked over the gifts to be handed out during the evening. Over the years, prizes of all kinds brightened gatherings. Many of the items given away were donated by ceramic companies or had ceramic components.

case are of such general use, in so many branches of our art, that members of all divisions should profit from more general diffusion of knowledge concerning them. So we should one and all welcome our new Division into our circle and encourage its members to do their best to represent worthily their field, with full confidence that they will help us more than we can help them.

And, as usual, Orton was correct. The division has been a vital and consistent part of the Society's programming and publishing efforts, participating in the annual meeting program and also holding an annual fall meeting, usually with the Whitewares Division. As a longtime participant put it, "You can't make whitewares without having materials."

For years these fall meetings were held at Bedford Springs. More recently, the division has held its fall meetings as part of the Society's Biennial Workshop program. Proceedings from the division meetings are published as part of the *Ceramic Engineering and Science Proceedings* series.

The division members and the companies they represent became active supporters and participants as the Society's exposition program grew, and today participants represent a significant portion of the Society's exhibitors and advertisers.

Although the growth of the nuclear industry had been dramatic in the 20 years after the atom bomb, the Nuclear Division of the Society was not formed until 1965, mostly because Society leadership was concerned that there were already a sufficient number of divisions and a new one might drain membership from existing groups. The focus of the division at the end of the 20th century is environmental cleanup at defense and weapons facilities in the United States, Europe and Canada.

NUCLEAR AND ENVIRONMENTAL TECHNOLOGY DIVISION

Though first proposed in 1959 when 37 Society members signed a petition on its behalf, the Nuclear Division did not become a reality until 1965, more than 20 years after the birth of the atomic age. Basically, it ran afoul of a belief at headquarters that there already were too many divisions, but from 1959 to 1965, a technical programming effort prepared for the new division.

Most of the members drawn to the fledgling Nuclear Division came from the Refractories and Basic Science divisions. "Some of the traditional Refractories Division members may even be relieved to see us go," wrote an early division member, "since we may have inadvertently caused them to sponsor annually the symposium on nuclear ceramics, for which we continue to be grateful and appreciative."

Ironically, no sooner had the division been established than the industry witnessed a decrease in government research and development, resulting in fewer papers and stagnant growth in membership. To combat this the division instituted a "member-get-a-member" recruitment program that swelled the division ranks by 10 percent.

Even so, in the early 1980s the division was hit by a slow economy, funding cutbacks and layoffs at many nuclear sites.

The end of the Cold War brought the Nuclear Division new opportunities as environmental cleanup became a priority at defense and weapons facilities in the United States, Europe and Canada. Reflecting this expanded mission and scope, a 1993 proposal to change the division's name to the Nuclear and Environmental Technology (NET) Division was approved by the Board in 1994.

As the 1990s close, environmental regulations created by the Environmental Protection Agency and other government agencies continue to create opportunities for the NET Division. Whether working independently or as a Society liaison with other organizations such as the Federation of Materials Science, the NET Division contributes its unique background and expertise to the area of environmental remediation.

Environmental concerns affect ceramics at many levels, in research and manufacturing, and in products for remediation and containment. In 1994 the Nuclear Division became the Nuclear and Environmental Technology Division to reflect "initiatives in environmental remediation throughout the Department of Energy (DOE) complex," according to Carol Jantzen, who led the name-change effort. "It was appropriate because of the increased awareness of technology transfer between the DOE complex and industry."

REFRACTORY CERAMICS DIVISION

One of the original divisions created in 1919, the Refractory Ceramics Division continues as the major technology-transfer organization serving the refractories industry.

Its purpose, according to a 1988 mission statement is:

> To serve producers and users of refractories and high-temperature ceramics by providing scientific, technical and other information on the science, production and application of ceramics for use at elevated temperatures and hostile environments and by promoting the professional development of its members.

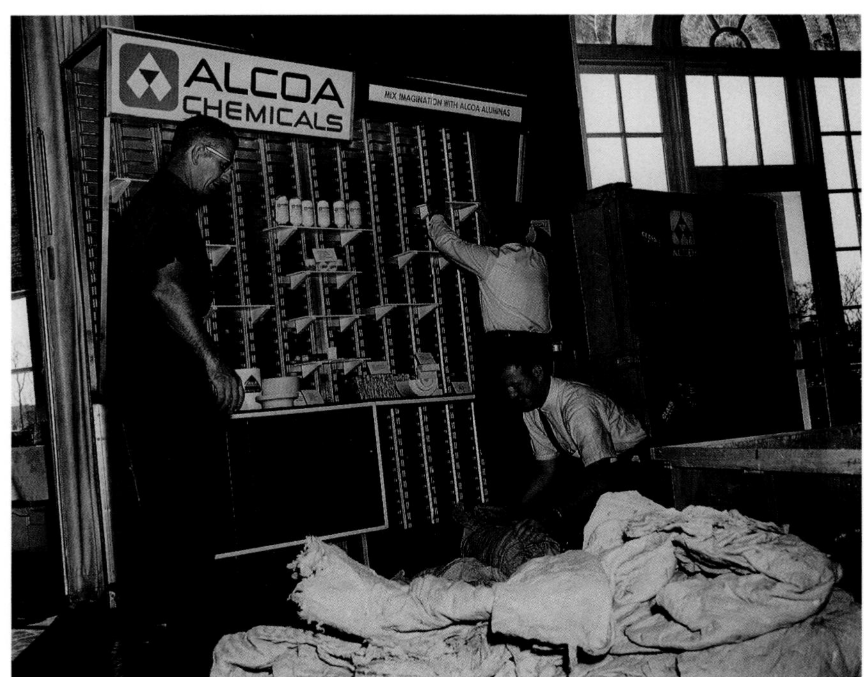

Exhibits, such as this ALCOA booth at the 1964 Refractories Division meeting, are a chance for ceramics companies and other allied industry concerns to reach Society members who may or may not be able to attend the annual meeting exposition.

As fitness consciousness took over the nation, the leisure activities of meetings expanded from traditional pursuits such as golf, fishing and trap shooting to include "fun runs," like this one at a 1982 Refractories Division meeting at Bedford Springs.

Divisions formed in the first quarter of the century reflected industry of the time — largely clay-based manufacturing. In the latter half of the century, new processes and products required divisions that wanted to keep their membership to change their names and arenas of inquiry, and new divisions were formed as well.

For many years, Refractories was the Society's biggest division, making up one-third of the entire organization's membership. In 1972 it was surpassed by the Glass Division. Today the Basic Science and the Engineering Ceramics Divisions account for approximately 50 percent of Society membership.

By the late 1980s membership in the Division plateaued, largely due to conditions in the U.S. steel industry and the technical advances in refractories that significantly extended their useful lifetimes.

Spawned by international conferences in Japan during the 1980s, the Unified International Technical Conference on Refractories (UNITECR) — an organization similar to the International Commission on Glass, rejuvenated the Division and created international interest.

The Refractory Ceramics Division, under the leadership of George MacZura, Dick Bradt, Charles Semler, Bob Fisher, Bill McCracken and others, played a key role in organizing the groups in Japan, Europe and South America that would form UNITECR, which held its first meeting in Anaheim, California, in 1989.

In 1997, the group's second meeting was held in New Orleans. Organized by the Society and the Refractory Ceramics Division, this meeting drew more than 1,200 attendees, more than half of whom had come from outside the United States. Future UNITECR meetings will rotate among the four principal groups, with the next one scheduled for Berlin in 1999.

STRUCTURAL CLAY PRODUCTS DIVISION

The history of this division is one of the most interesting and varied of all of the Society's divisions and classes. Its roots lie in the creation of the Terra Cotta Division in 1919 and the Heavy Clay Products Division in 1922. These two groups merged in 1940. Meanwhile, the National Brick Manufacturers' Association, at whose 1898 meeting the Society was first conceived, and which had existed for many years outside of the Society, merged in 1931 with the Heavy Clay Products Division to form the Structural Clay Products Division. During this same period the Society took over the management of the National Brick

Manufacturers' Research Foundation. In 1939 the Society terminated the operation because of financial difficulties.

By the mid-1960s, in fact, division leaders were lamenting the number and quality of papers presented by the division, a situation due largely to a general lack of research supported by companies within the industry.

"What research is being done is primarily oriented toward processing techniques and equipment, or hardware development," H. Howard Lund, then chairman of the division, wrote in 1967. "Unfortunately, such research that is being done is usually classified as 'Industrial Confidential' and is not made available for presentation. This veil of secrecy is usually only a figment of the imagination since secrets are most difficult to conceal for any length of time."

Spearheaded by a core of dedicated officers, this Division broke with tradition during the 1990s and began organizing its fall meetings in the spring, a move that met with dramatic success.

B.H. Drakenfeld and Company, Inc. (today Cerdec) displayed a broad selection of products at the 1948 Annual Meeting to show that the company that manufactured color and chemicals for the glass and ceramics industry had surpassed its pre-war output. Beginning in the late 1930s, delegates to the Society's annual meeting each received a commemorative Drakenfeld glass tumbler; the joke was that even if the meeting didn't look promising, no one wanted to miss getting the year's Drakenfeld glass.

WHITEWARES DIVISION

One of the second series of divisions created by the Society in 1921, Whitewares represents traditional ceramics, including earthenware, semivitreous ware, porcelains and chinas used as artware, dinnerware, floor and wall tiles, sanitary ware, chemical ware, high- and low-voltage porcelains and special thermal and mechanical wares.

At various times over the years it has been suggested that Whitewares also claim certain new electrical and technical ceramics: alumina, beryllia, forsterite ceramics, ferrites, cermets, titanates, zirconates, nuclear fuels and foamed ceramics.

Under the broadest definition of "whitewares," some of these ceramics could conceivably have found a home in the division. But they also were closely related to the sciences and products represented by numerous other divisions, such as the Nuclear Division, which only came into existence relatively late in the Society's history.

In any case, the Society's approval of multiple divisional memberships made this a moot point, allowing ceramists to follow their interests in particular disciplines into several divisions.

One of the first divisions was the Terra Cotta Division, which eventually merged with the Heavy Clay Products Division, as terra cotta fell out of favor as a building material.

FI

If this isn't miss shotgun of 1981 I'll eat six clay pigeons.

You're looking good sweetie

cheers

Reg

Roger Jones

FIRELINE, Inc. Jones Street Youngstown, Ohio 44502

High Temperature Insulating Shapes - and - Thermal Shock Resistant

To new members, stories of afternoons spent trap shooting may seem the stuff of fable, but recollections suggest clay pigeons were almost as important as technical papers to some division meetings. Jo Gitter was still perfecting her aim during the Refractories meeting in 1981.

Today the Whitewares Division continues to represent traditional whitewares. Though in recent years suggestions have been raised to change its name to reflect more accurately its concerns (many product lines represented by the division cannot even be considered "white"), it retains the same name it had when it came into being 80 years ago.

Divisions are the main organizational structure for members' varied interests. But other structures, such as sections and classes, also channel members' energies providing good working relationships, friendships and opportunities to share professional growth. Classes within the Society — the National Institute of Ceramic Engineers, the Ceramic Educational Council and the Ceramic Manufacturing Council — also focus members' interests and efforts on particular areas within the arena of ceramics. While some of them are as large as professional associations in other industries, they continue to operate within the Society, maintaining the connections, information exchange and fellowship that have marked the association of ceramists for a century. ▲

In the late 19th century, factory work in ceramics was often in whitewares, which at that time were largely earthenware, dinnerware, artware, floor and wall tiles and sanitary ware. Later, whitewares came to include other products such as high- and low-voltage porcelains, electrical and technical ceramics, forsterite ceramics, ferrites, cermets, titanates, zirconates, nuclear fuels, foamed ceramics and others. Name changes have been regularly discussed to accommodate this expansion. Ralston Russell, Jr. in the mid-1960s addressed the issue by advocating "Fine Ceramics" for ceramics of fine microstructure.

AN INTERNATIONAL FLAIR

Reciprocal relationships with ceramists around the world have been valued in the Society since the early 1920s. By the 1980s, the availability of affordable international travel meant more personal visits were possible between members and hosts or guests from all over the globe.

From the occasion of the Chinese visit in 1980 forward, international travel became the order of the day. Only one year later, in 1981, the Society took part in the 90th Annual Meeting of the Ceramic Society of Japan. The trip was an opportunity to see first-hand the industrial companies and government and university laboratories, and technical efforts and accomplishments of the Japanese. It also afforded the American delegation the opportunity to attend the opening of the Tokyo Institute of Technology Centennial.

During their visit Society delegates were recognized by their colleagues from three continents as members of the world's leading scientific and engineering organization in the field of ceramics. Such recognition affirmed that the newly formed International Ceramic Society Coordinating Committee was helping the Society achieve its goal of building a more formidable international presence.

In 1984 a formal delegation from the Society attended the Annual Meeting of the Ceramic Society of Japan and acquainted Society members with Japan's "ceramic fever," which was evident in its commitment to the industry. Society members were impressed with Japan's long-term research and development sponsorships that showed what then-President Richard Spriggs called ". . . wide-spread industrial commitment of manpower and resources to high technology ceramics . . ."

Other international visits and technical exchanges included those to the annual meeting of The Brazilian Ceramics Association in 1982; the annual meeting of the Ceramic Society of Japan in 1984; the New Materials '86 Exhibit in Hanover, Germany; the General Electric Research and Ecole National Superieure des Mines de Paris in Paris, also in 1986; Korea's KCerS annual meeting in 1987; the official signing ceremony forming the European Ceramic Society held in England in 1987; Austceram in Sydney, Australia, in 1988; the celebration of the Ceramic Society of Japan's centennial in 1991; and

All aboard The American Ceramic Society Special for world ports of call! A 1928 trip to Europe set a standard for international exchange with ceramists around the world.

This photograph was titled by its presenters, the Blythe Colour Works, Ltd. of Crosswell, Stoke-on-Trent, England, as "An incident in the visit of The American Ceramic Society to Stoke-on-Trent, July 1928." The American delegation went to England to observe the 200th birthday of Josiah Wedgwood.

Delegates toured West Berlin in 1986 as part of an ACerS exchange visit that included France and Germany in conjunction with New Materials '86, called the "Hanover Fair." The exchange included technical presentations, visits to universities, industries and government laboratories, as well as sightseeing in both countries.

(Left) A delegation of Society members and spouses to Korea in 1987 enjoyed a picnic banquet in a garden south of Seoul, Korea.

Around the world, ceramics companies such as the Lamosa Tile Plant in Monterrey, Mexico, (above) and the Asia Cement Corp. Hulien Plant, Taiwan, (right) benefit from the information and collegial exchange the Society offers.

the 50th anniversary celebration of the Chinese Ceramic Society in Beijing in 1995.

The Society has also recently signed affiliation agreements with the Chinese Ceramic Society, the Mexican Ceramic Society and the Ceramic Society of Japan, in each case committing the two organizations to working closely together to enhance information exchange through publishing, education and programming endeavors.

As American ceramists traveled abroad to learn more about their industry, their contemporaries also traveled here. The United States hosted the First Unified International Technical Conference on Refractories in 1989 concurrently with the Society's First International Ceramic Science and Technology Congress.

In the Society's international exchanges, whenever possible, the planning committee coordinates visits with the meetings of other organizations, enabling guests to participate in the formal technical programs of the meetings. Many visits have been followed by reciprocal host visits, such as the Korean delegates' visit to the ACerS Annual Meeting in 1988.

These numerous visits have provided Society members with important glimpses of ceramics' roles in other countries. They also have historically been seen as a way to establish common standards, form consistent terms of reference for economic data, improve data in published material, improve timeliness of journals and patents and raise interest in new ventures. ▲

William H. Payne (left), ACerS president and George MacZura, president of UNITECR '89, shook hands over a job well done when the first Unified International Technical Conference on Refractories (UNITECR) attracted delegates from around the world to Disneyland in Anaheim, California, in November 1989. The American Ceramic Society was joined in the effort by the German Refractories Association, Association of Latin American Refractory Manufacturers, and the Technical Association of Refractories-Japan.

(Above) In 1997, the Mexican Ceramic Society, Northern Chapter, dedicated its convention to Malcolm G. McLaren — longtime ACerS member and champion of international exchange — and announced that the opening address of the Society's annual meetings would henceforth be known each year as "The Malcolm G. McLaren Keynote Address." Mrs. Malcolm McLaren (Barbara) accepted a certificate from Mario Zamudio, chapter president.

(Left) One of the benefits of international exchange visits is the opportunity to experience the social side of meetings in other countries, as did the 1984 delegation to Japan.

Technical and cultural exchanges between the Korean and American ceramic societies during 1987 and 1988 followed years of official planning meetings for officers and directors of both groups, building on relationships established when Korean colleagues came to the United States after the Korean War to pursue graduate studies in ceramics. The exchange program began in 1987 with a trip to the KCerS fall meeting in Seoul and concluded in May 1988 when the Korean delegates attended the ACerS Annual Meeting in Cincinnati.

At the 1997 Annual Meeting, the Society looked ahead to a new century of international cooperation when ACerS President Carol Jantzen and Naohiro Soga, president-elect of the Ceramic Society of Japan, signed a proclamation for a partnership effort between the two societies.

Responding to a formal invitation, a Society delegation, shown here in front of Chinese Building materials Academy headquarters, included Victor Greenhut, Gong Fangtian, Delbert Day, James McCauley and Zheng Yuan Shan. They participated in the 50th anniversary activities of the Chinese Ceramic Society in the Beijing area in October 1995. The agenda included a formal plenary session and luncheon; tours to several ceramic plants as well as educational and research institutions; discussions with senior scientists and engineers; and a farewell banquet with senior dignitaries.

JOURNAL
OF THE
AMERICAN CERAMIC
SOCIETY
VOL. 1
1918
ABSTRACTS
AND
BULLETINS
VOL. 1
1922
RECORDS
AMERICAN CERAMIC SOCIETY
STATE OF OHIO
TRANSACTIONS
OF THE
AMERICAN CERAMIC SOCIETY
CONTAINING THE PAPERS AND DISCUSSIONS
OF THE
FIRST ANNUAL MEETING
HELD AT
COLUMBUS, OHIO
FEBRUARY

GLOBAL PUBLICATIONS

The American Ceramic Society will be 50 years old in 1948. It is time for it to shed its baby clothes and start to wear long trousers. There is no reason why nearly every paper for the annual meeting should not be presented far enough in advance so that it will pass through the Editorial Committee and be preprinted before the annual meeting. • All the other activities of the Society are subordinate to and complementary to the one vital function of the publication of ceramic literature. • So wrote President J.D. Sullivan in 1947, and his observation has stood the test of time. At that time, a study revealed that 80 percent of the Society's activities were connected to publishing. Today publishing constitutes the single largest operation of The American Ceramic Society, making accessible ceramic literature being produced by the membership and ceramists around the world. • Among the first acts of the newly founded Society was to authorize two land-mark publishing ventures. • The first was the creation of an 80-page pamphlet, *The Manual of Ceramic Calculation,* which was viewed as a means of assisting ceramists to advance their work from an empirical to a scientific basis.

(Above) Three staff members who worked on publications in the 2525 North High Street office in 1936 sat in a room that combined the necessities of publishing with the Society's growing collection of ceramic objects.

(Opposite) Even before it had much clout as a membership organization, the Society was a powerful voice in ceramic science through its publication of papers and books.

The Journal of the American Ceramic Society *has been a collaboration of volunteer editors, member contributors and staff (at first university staff where editors were teaching then, after 1922, paid Society staff). In 1954, Herbert Insley (center) met with staffers Mary Gibb (lower left), who edited more than 170,000 abstracts during her 41-year career with the Society; Margie K. Reser (second from left), technical editor for 21 years; Ginny Benedict and Ms. Weigelt. Insley served as editor of Society publications from 1954 to 1972.*

Edited by Charles Fergus Binns and first published in 1900, the manual became one of the Society's perennial best-sellers. This publication was made available for $1 so manufacturers could afford it and was printed on sturdy paper in a format that would hold up on the factory floor.

The second was an English translation of *The Collected Writings of Hermann Seger*, the famed ceramist at the Royal Porcelain Factory in Berlin who dominated research in the field in the late 1800s. Edited by Albert V. Bleininger, the Society's volume became the first principal source of scientific literature available in English to American ceramists.

That volume was described by Society President Francis W. Walker as:

> the monument that will stand to our greatest credit with the English-speaking world . . . This work was undertaken at the organization meeting of our society, and after five years of laborious work on the part of the editor, the committee in charge and some of the members, they were able before the close of the fifth year to put within the reach of all English-speaking people this invaluable document.

A third publishing venture was *A Bibliography of Clays and the Ceramic Art* compiled by John C. Branner of Stanford University. The volume provided references to more than 6,000 titles, about twice the number available in a previous edition published in 1896 by the U.S. Geological Survey.

But it sometimes took years to produce a book, and early Society members wanted a publication that would give them the latest information regarding science and research in the ceramic field.

THE *TRANSACTIONS* AND THE *JOURNAL*

Founding members of the Society voted to publish a quarterly volume (which after a year became annual) called *Transactions of the American Ceramic Society*. This journal would allow for the presentation of papers and scientific news. The first volume covered the activities for the year 1899; it contained eight papers and five discussions on 110 pages.

"You can get away with anything around here the day the Ceramic Bulletin comes in."

In 1961, a cartoon poked fun at the absorbing material in The American Ceramic Society Bulletin.

From the Journal*'s first advertising solicitation, a letter from A.V. Bleininger, publications committee chairman:*

. . .I am personally appealing to your interest in the American Ceramic Society, which numbers representatives of your firm in its membership, to support this Journal in its initial year. "The Journal of the American Ceramic Society": will be the only periodical that will interest and eventually reach every engineer, chemist, factory manager, shop man and owner of every branch of the Ceramic Industry. . . The subscription price is to be $6 annually . . . The advertising rates for the first year are as follows: Full page—$40, Half—$25, Quarter—$15, Cover pages-$50 (Discount of 20% allowed for contracts of 12 continuous insertions). . . I sincerely trust that you will consider this arrangement in a sympathetic spirit and make effective use of this advertising opportunity for the twelve numbers of 1918.

The Society's founders were determined that the *Transactions* become as comprehensive a record of the current ceramic scene as possible. They demanded that:

> All papers, discussions and other writings which have been presented before the Society, its Divisions or Sections, and all Committee Reports shall become the property of the Society and shall be transmitted to the Editor immediately after presentation.

Today the Society makes no pretense of having legal rights to every paper read at one of its functions.

Initially *Transactions* was to contain all papers presented at the annual meeting. The first departure from this policy came in 1902 when several papers that had not been read before the Society were published.

It immediately became obvious that the *Transactions* filled a void in the world of technical literature. In 1902 General Secretary Edward Orton, Jr. reported a request from a Peruvian ceramic engineer for the third volume of the *Transactions*.

"Similar requests from other portions of the globe would indicate that the Society's field of usefulness is almost without bounds," observed *The Clay-Worker*, a commercial periodical covering the clay industry.

By 1918 *Transactions* had reached a size of ludicrous proportions — one volume reportedly was seven inches thick. At that time it was decided to publish monthly, and the name was changed to the *Journal of the American Ceramic Society.*

The *Journal* was devoted to "the arts and sciences related to the ceramic and silicate industries, a technological publication which contains records of original research."

The Society's members were determined that the *Journal* become the premiere scientific publication on ceramics — a position it has held without question for 80 years. To ensure quality, the editors went so far as to develop a policy for those papers that would NOT be published in the *Journal*: those containing matters readily found elsewhere, those specially advocating personal interests, those carelessly prepared or controverting established facts, and those purely speculative or foreign to the purposes of the Society.

Among the early editors was Ross C. Purdy, whose service proved an able apprenticeship for his long reign over the Society's publishing enterprises in the 1920s and 1930s.

Even in the early years of the *Transactions* something of a rivalry had existed between what one ACerS historian termed "the practical man and the scientific man, between whom now stands the so-called technical man."

The question of the emphasis of the Society's publications — whether to stick to pure science or to take a more pragmatic approach toward engineering and problem-solving in the plant — has been debated within the Society's ranks for a century. Only in recent years have specific publications been developed to address the needs of a membership that ranges from pure scientists to engineers to managers and sales personnel.

In 1909 the Society's president made a plea for more practical papers, noting that with increased knowledge and education among members, papers of "an abstruse nature" geared to the student and researcher were dominating the publication.

> It is essential that each number of the *Journal* shall contain original papers or other material of practical value to each of the industries represented by our membership. It is hoped that the full development of the system of establishing local sections and professional divisions will result in the contribution of more papers, especially of the class designated as practical, since they deal directly with factory problems.

The worldwide acceptance of the *Journal* has been demonstrated by its foreign circulation. By 1950 approximately 1,000 copies of each issue went to 49 different countries. Today, the *Journal* goes to 56 countries through the mail and, in addition, the Society's World Wide Web site publishes titles, authors and abstracts for all papers published since January 1997, available to anyone who wants to keep up with scientific work in ceramics.

THE *BULLETIN*

The practical-vs.-scientific schism was addressed with the creation in 1922 of *The American Ceramic Society Bulletin,* first as a separate publication and then under the same cover as the *Journal.*

Thus the *Journal* dealt largely with scientific papers of permanent value, while the *Bulletin* provided an outlet for timely news of particular interest to "practical and technical men." It was described as "a forum of mutual help for members who were concerned with production and plant problems, discussions of technical and scientific questions and promotion of cooperative research."

In addition, "The *Bulletin* will not compete with

ON THE LIGHTER SIDE

Even serious ceramic engineers and scientists occasionally need a chuckle. Such was the motivation behind an ad for a cartoonist in the March 1963 issue of the Bulletin. *Cartoons added a touch of levity to Society publications by revealing ceramists' human sides. For this reason, "practicing ceramists" with a feel for the industry were encouraged to submit their cartoons or ideas for cartoonists.*

By June of the same year the Bulletin *reported that judging from the response to the March ad, "ceramists do have a delightful sense of humor." One particular cartoonist, who went by the pen name "Jo Ro," submitted entries with the "rare combination of talents that is just what the editors have been longing for but didn't expect to uncover." The editors praised all of the entries they received and looked forward to more offerings from the Society's versatile and talented members.*

Indeed, cartoons punctuated the pages of the Bulletin *for years, penned by ceramists from all areas of interest. Cartoons have provided a lasting record of the humor that has always marked Society gatherings. Most witticisms were merriment of the moment saved only in recollections of members as the lighter side of some papers and meetings.* ▲

AUDIENCE

A group of people standing, sitting or slumping in front of the speaker. They have come to hear the paper just prior to or immediately after the one being given.

nor to any large extent serve the same purposes as the trade journals. General news items and plant descriptions will not be featured in these columns."

It was on the pages of the *Bulletin* that Ross Purdy, the Society's first paid secretary (and also its editor-in-chief) wrote countless editorials in an effort to nudge, browbeat and convince the membership to move in the direction he deemed best for the organization he loved.

The *Bulletin* also served another purpose of great importance to the Society's survival — it alone among the organization's publications sold advertising. By 1962 earnings from advertising in the *Bulletin* accounted for 40 percent of the Society's income (dues and subscriptions accounted for 45 percent).

In 1933 the *Bulletin* was separated from the *Journal*, allowing it to be distributed to nonmembers and others who were interested in the ceramic trade but had little interest in the scientific discussions contained in the *Journal*.

Just what goes in the *Bulletin* and what in the *Journal* has been a matter of much debate over the years. In 1946 the Publications Committee tried to differentiate between the approaches of the two publications: The *Journal*, they ruled, would contain all papers of original work, whether or not of highly scientific nature, just as long as the data are complete and accurate. The *Bulletin* would contain papers on application or reiteration of existing information of sufficient worth. In such cases scientific papers could be included in the *Bulletin* but "should be written in a popularized and not technical form."

Basically, where a paper was published was based strictly on its scientific content. All scientific papers, whether original or reiteration, should be published in the *Journal* and only papers of application and popular nature would be presented in the *Bulletin*.

Over the years the *Bulletin* has undergone several graphic face-lifts and editorial makeovers — among the most significant was an effort in the early 1950s to make the *Bulletin* more interesting to plant personnel "so that they would have some reason for being interested in the Society if they were not of a scientific turn of mind."

In 1950 the *Bulletin* was admitted to the Audit Bureau of Circulation, an organization certifying the fidelity of the circulation of publications for the advertising profession, a status highly regarded in the advertising field. At the same time Charles Sidney Pearce, then the Society's General Secretary and the

Bulletin's editor, was elected to membership in the Society of Business Magazine Editors.

"It would seem," Pearce declared, "that these two connections now prove that the *Bulletin* is no longer a publication of 'amateur adventurers' but that it has assumed full stature in the field of business papers."

For years there was discussion of making the *Bulletin* a magazine of more popular appeal, one that would interest the operating men in the industry much more than the scientific and technical men.

In 1982 the *Bulletin* was redesigned to include more member news, more stories on business and management trends and more special issues. "Even the technical papers are intended to become engineering-oriented," President Robert J. Beals informed the membership.

Ten years later another redesign created sections on manufacturing, engineering and technology, plus one stand-alone paper showcased in each issue as the cover story. The reorganization reflected the goal of better serving manufacturing, end-user and member groups.

In 1994, Patricia A. Janeway was named editor of the *Bulletin*. An avid advocate of industry causes and campaigns, Janeway brought a broad perspective about ceramics to the *Bulletin* that increased the manufacturing focus, while continuing the coverage of engineering, technology and Society news. Today the *Bulletin* continues to be the industry's premiere forum for ceramic news in diverse areas of interest.

Since its inception, one of the Society's major responsibilities has been to facilitate the exchange of information among its members. Serving as a companion to the *Bulletin* is a directory that is a ceramic resource guide. In 1985, *ceramicSOURCE* was introduced to provide an annual, comprehensive directory of company, product and services information. The annual American Ceramic Society Roster that includes members' mailing addresses, phone and fax numbers, and e-mail addresses also aids communication.

Keeping members in touch around the world in 1998 is the Society's Web page at *www.acers.org*. The information superhighway holds great promise for the Society as it begins a second century of connecting members to information they can use. The Society continues to create new on-line identities for all its publications, to ensure that all information that it publishes is available at the touch of a button.

General Secretary Charles S. Pearce was also editor of the Bulletin *and was determined that it be regarded as more than a "publication of 'amateur adventurers.'"*

Francis Flint (head of table, right), Society president in 1936-1937, held a publications committee meeting in Columbus with staff members, including Emily Van Shoick (left), assistant editor of the Journal, *who served on the publications staff of the Society for 26 years.*

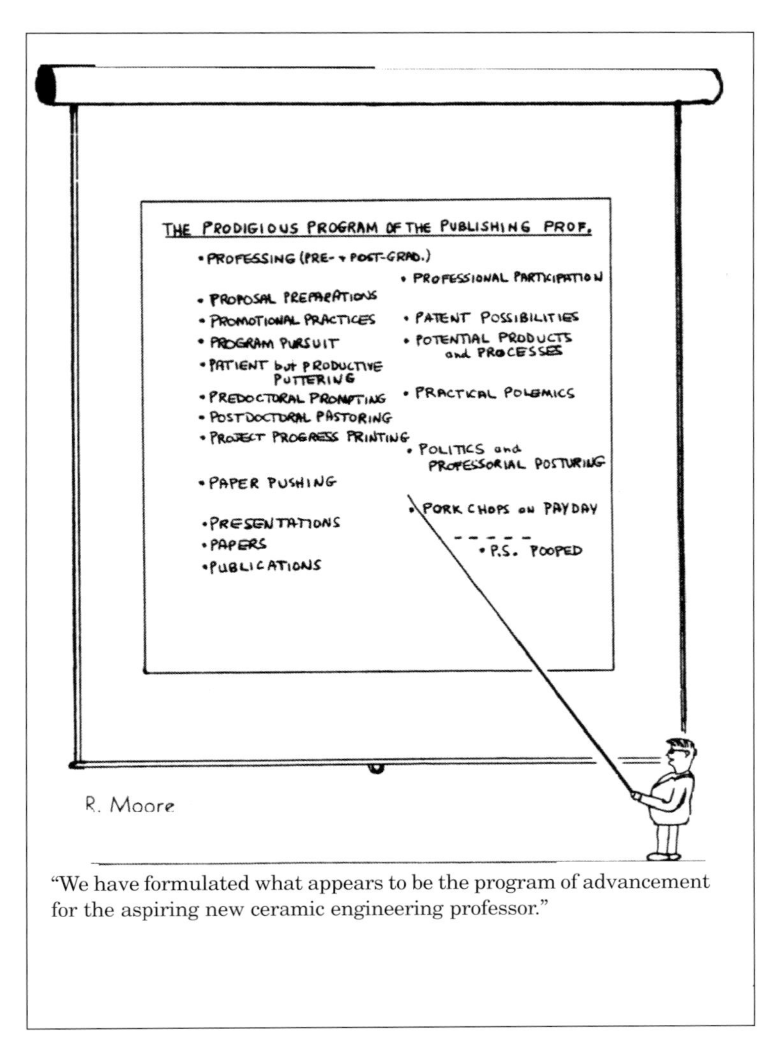

"We have formulated what appears to be the program of advancement for the aspiring new ceramic engineering professor."

A cartoon contributed to the Bulletin *in 1968 made light of the important role scholarly and technical publications such as the ones put out by the Society can have in some careers in ceramics.*

CERAMIC ABSTRACTS

As early as 1904 the Committee on Ceramic Literature under Charles F. Binns was working on a digest of notable recent additions to ceramic literature. This project, which ultimately resulted in *Ceramic Abstracts*, reviewed the work of the German, French and English technical press as well as that of ceramic societies around the world.

The first abstracts appeared as part of the *Journal* beginning with Volume 2 in 1919. *Ceramic Abstracts* became a publication in its own right in 1922, offering abridgments of articles of direct interest to the ceramic industries. The first six volumes were published under cover with the new *Bulletin*. *Ceramic Abstracts* was published under separate cover beginning with Volume 7 in 1928. For a brief time, 1962 through 1974, *Ceramic Abstracts* was published with the *Journal*.

The task of reviewing a world's worth of literature and condensing it to a well-organized and useful collection of abstracts has always been a time-consuming and demanding enterprise. Several times during its history the Society has seriously reviewed the resources — time and money — demanded by this undertaking and questioned whether it should continue the enterprise.

In the mid-1990s the Society made a commitment to evaluate seriously this part of its publishing program. It became apparent that, while the service of providing ceramic abstracts was a very important one, the Society did not need to administer the program directly. Thus, in 1997, the *Ceramic Abstracts* program was sold to a commercial abstracting service.

PHASE DIAGRAMS

Among the most ambitious publishing efforts of the Society has been the *Phase Diagrams for Ceramists* program. Considered by many members as one of the most important contributions of the Society to ceramists worldwide, the phase diagram program is a clear example of the way ceramic research has aided people working in factories and other applied disciplines.

Phase diagrams are used extensively in the research and development of technologically important materials because they provide guides to how materials will respond to different temperatures. Because all ceramics involve heat, there are very few ceramic inquiries that do not make use of phase diagrams. The *Phase Diagrams for Ceramists* was first published in the October 1933 issue of the *Journal*, edited by F.D. Hall and Herbert Insley. Since then, the *Phase*

Diagrams for Ceramists project has grown into an international effort through the enlistment of a worldwide team of contributing editors, experts in various fields of ceramic phase equilibria.

For 40 years volumes of the diagrams were published as they were completed. But the volume of research in the field was so great that by 1981 the publishing efforts had fallen at least five or six years behind the new information. The Society partnered with the National Bureau of Standards (now the National Institute of Standards and Technology) in a joint effort to make the program current and to develop a data base of phase-related information.

A fundraising effort secured nearly $2.5 million in pledges from corporations and individuals, allowing the program to sustain its effort. In 1992 President Dennis W. Readey announced that the *Phase Equilibria* publications were now available as a personal computer database. By making the phase diagram so readily accessible, ACerS had achieved all of the goals of the program begun in 1933.

AMERICAN CERAMIC SOCIETY COMMUNICATIONS

In 1980 the Board approved establishment of a new monthly journal, the *American Ceramic Society Communications* to be published with the *Journal*. It contains brief communications of scientific and technical interest, important new conclusions that will be described in a later paper, discussions of new hypotheses based on theory or experimental evidence and significant comments on publications relevant to ceramic science or engineering.

The primary purpose of this enterprise is to provide for fast publication of short papers.

ADVANCED CERAMIC MATERIALS

In 1986, the Society introduced a new quarterly publication devoted exclusively to high technology ceramics for electronic, electrical, mechanical, magnetic, and optical applications. *Advanced Ceramic Materials* included new releases and overview articles as well as peer-reviewed papers. In 1987, at the direction of the Publications Committee, the frequency was increased to bimonthly.

In 1988, significant changes were made in the management of the Society's peer-review process, including the introduction of Journal Editors, respected members of the ceramic community chosen to oversee and adjudicate the entire peer-review

DON'T JUST STAND THERE . . . WRITE!

Only about 10 percent of our membership contribute papers in any one year. About half of these are regular contributors. Five percent of the membership has done 70 percent of the work. This is deplorable. Every member should feel it his duty to present a paper or a short note on some subject at least every other year. Every man in the ceramic industry has opportunity to collect sufficient data to do this.

The real explanation is the refusal of the majority of our members to make the effort necessary to write up the data in their possession. They seem oblivious to the fact that the men who do prepare papers are spending time and labor for their benefit.

— *editorial in the* Bulletin, *1925*

In the early 1950s, equipment used to put out Society publications didn't look much different than it had since the Society consolidated administrative and editorial functions in 1922 (previously the all-volunteer leadership meant that editorial work was done in Illinois, where editor Charles Binns was then and administrative work was done by Ross Purdy in Ohio). A typewriter, sharp pencils, glue and scissors served the tasks that today have expanded to require sophisticated computer hardware and software.

Luther E. Ohrstedt (second from right), shown here in 1952, started on the Bulletin *as a journalism student in1946, left the Society in 1952, returned in 1968 and until his retirement in 1982 served as advertising director.*

Publications of The American Ceramic Society are at once a lively current exchange and a legacy to future ceramists.

process for journal-type papers. Robert E. Newnham and Arthur H. Heuer were the first members chosen to hold this office.

As part of this overhaul, it was decided to incorporate *Advanced Ceramic Materials* into the *Journal*, thus bringing the entire review and publication process under one umbrella. President William H. Rhodes and Society Editor J.B. Wachtman, Jr. announced these changes in the September 1988 issue of the *Bulletin*, and the last stand-alone issue of *Advanced Ceramic Materials* was published in December 1988.

CERAMIC ENGINEERING AND SCIENCE PROCEEDINGS

Society Editor William Smothers and Executive Director Arthur Friedberg were keenly aware of the practicing ceramists' need for collections of practical papers—papers that did not need to pass through the stringent review process typified by the *Journal* before publication, but rather papers that contained kernels of useful, applicable information and, thus, deserved to reach a wider audience. They also knew that papers like this were being presented annually at topical meetings such as the Porcelain Enamel Institute meeting, Structural Clay Products, the combined Whitewares and Material and Equipment divisions' meetings, the Automotive Materials Conference, and annual conferences such as the Engineering Ceramics Division's Cocoa Beach Conference.

Smothers proposed that the Society create a new publication specifically designed to gather, publish and distribute these proceedings in a timely manner. Thus, in 1980 *Ceramic Engineering and Science Proceedings* was added to the Society's growing list of publication enterprises. Today, CESP continues to meet the needs of this audience and, with the use of modern word processing technology, is known as a vehicle for prompt dissemination of meeting papers.

ADVANCES IN CERAMICS

In 1980, a new book series, *Advances in Ceramics*, was approved. In 1979, Executive Director Arthur Friedberg recommended to the Publications Committee that a policy be established to involve the Society in the publication of a wide variety of material relating to ceramic engineering and technology to include conferences and proceedings. Each volume would represent a collection of papers gathered and peer-reviewed at the direction of the volume editor, who was usually the meeting organizer as well. The

A "FAIRLY GOOD IDEA" REIGNS FOR EIGHT DECADES

The American Ceramic Society's original seal cost $15 to design. It turned out to be money well spent, since the logo was used for nearly 86 years.

In the October 1901 issue of Brick, *the American Ceramic Society issued a call to its members to design its seal. The prize for winning was to be a two-volume set of* The Collected Writings of Hermann Seger, *valued at $15.*

Competition was brisk, and a committee carefully studied the entries, hoping to find a design that would symbolize the eventual scope of the fledgling organization — and would look good on a membership pin. It turned out to be difficult to find an appealing combination of substance and style.

R. Guastavino, Jr., whose company specialized in "fire-proof construction" in New York City, won the prize, despite the fact that his design apparently wasn't Society Secretary Edward Orton, Jr.'s original choice for first place. Although Orton liked Guastavino's design idea, his artwork apparently left something to be desired. Orton described the entry with this note: "I consider the fundamental idea fairly good, but the drawing and technique are poor."

Once Guastavino had made some artistic refinements, the Society board eventually chose his design, on the merits of its general idea. They liked its unique way of showing some of the ceramic industry's most common products in a not-so-common setting.

In the seal's center sits a classical female figure, representing Science. In one hand she holds a set of scales, to signify one of ceramics' most common tools. In the other, she holds a chemistry retort, to symbolize chemistry's role in ceramics. Surrounding Science are ceramic goods, ranging from crude products such as pipe and brick to fine art pieces like plates and vases. Behind this scene sits a kiln, because, as Guastavino said in his contest entry, "all clay products are trusted to its care for the final completion."

The seal showing Science and her ceramic products underwent a few modernizing overhauls through the following years — Science's formerly basic facial features became more refined (and more attractive by modern standards). A

cartoon series in one annual meeting program (above) lampooned the process.

In 1987, ACerS gave up the well-liked but rather worn classical image, replacing the 86-year-old seal with a more contemporary logo design. The new logo adds the representation of both the technology and the international community that have become such integral parts of the ceramics industry and of the Society in the twentieth century.

The new logo is made of three overlapping elements: a triangle, representing phase diagrams; a globe, symbolizing the Society's international membership; and a sweeping S, which symbolizes how the Society pulls together various aspects of the ceramics community. For the Centennial Celebration it has carried an additional date line.

As up-to-date as the new logo looks, some members miss Lady Science, queen of the Society's printed pieces for so long. It's hard to imagine her proud presence cost the organization only about 17 cents a year all her life. ▲

The American Ceramic Society

The American Ceramic Society
1898-1998
A CENTURY OF CERAMICS

first volume published was *Grain Boundary Phenomena in Electronic Ceramics*, edited by Lionel M. Levinson and David C. Hill. Approximately 30 volumes were published in this series.

The Publications Committee met in 1955 with staff in the new Ceramic Park building: (left to right) J.O. Everhart, OSU; Margie K. Reser, staff; J.J. Canfield, Armco Steel Corp.; W. Raymond Kerr, Armstrong Cork Co.; Clarence H. Hahner, National Bureau of Standards; Mary Gibb, Robert Ransom, Virginia Benedict, staff; and Charles Pearce, editor and general secretary.

CERAMIC TRANSACTIONS

While *Advances in Ceramics* provided collections of presented papers which were peer-reviewed and carefully reproduced, the technical community began to express a need for a more prompt and less expensive vehicle for the publication of proceedings. *Ceramic Transactions* was introduced in 1988 with Volume 1, *Ceramic Powder Science*, edited by Gary L. Messing, Edwin R. Fuller, Jr. and Hans Hausner. To date, more than 80 volumes have been published.

CERAMICS MONTHLY

In 1996 the Society acquired *Ceramics Monthly* magazine, a publication serving the ceramic artist and potter and the general design community. The fine arts end of ceramics has always meshed uneasily with the dominant scientific strain in the membership; acquisition of this magazine allowed the Society to better serve this important segment of the field.

Ironically, Spencer Davis, who founded the magazine in 1953, had been a member of the Society's headquarters staff for many years and had first proposed that the Society publish the title. He felt that the ceramic arts community was not being adequately served and that a magazine designed for this audience would be a success. Society officials, particularly General Secretary Charles Pearce, were not receptive. With the confidence of his convictions, Davis set out to publish the magazine on his own and achieved a circulation of approximately 5,000 within one year. When Davis sold *Ceramics Monthly* upon his retirement in 1996, the magazine's circulation approached 40,000.

At the 1997 Annual Meeting, a "birthday party" celebrated the 75th anniversary of The American Ceramic Society Bulletin. *Editor Patricia A. Janeway blew out the candles and made a wish on behalf of the Society for the* Bulletin *to continue its successful service to members worldwide.*

In the Centennial year, the staff and equipment of the publications department in the Society headquarters in Westerville, Ohio, reflect the current competitive environment of publications and graphic design, as The American Ceramics Society works hard to assure the continued value of its publications to members.

OTHER PUBLICATIONS

The first edition of *The Ceramic Glossary* was published in 1963. It contained more than 1,700 ceramic definitions.

"We anticipate there will be brickbats flying over some of the definitions," observed Emily Van Schoick, editor of the volume and for 26 years a member of the Society's editorial staff. "But this should be all to the good. With copies in the hands of many interested people we can hope for suggestions for additional terms, more definitions and refinements of those in this first edition. A start had to be made somewhere and this is it."

Previous glossaries in the ceramic field were limited to a single segment. The *Glass Glossary* was published in 1948, and *Enamel Glossary* in 1950 and a compilation of terms for the Whitewares Division appeared in 1952.

Other series of books published under the auspices of the Society include:

Ceramics and Civilization
Cements Research Progress
Materials Science of Concrete

as well as many significant stand-alone titles. ▲

Box 3, West Wareham, Mass
1902-6-5

Prof. Edward Orton, Jr.,
Columbus, Ohio.

Dear Sir:- Have just received my "Slag Constitution" paper from the printer. Its publication has been somewhat delayed. I send a copy under separate cover. If convenient, I should very much appreciate a copy of your paper "On the Production of an Easily Fusible Glass without Lead Oxide or Boracic Acid."

Yours truly,
Harrison Everett Ashley

Before The American Ceramic Society began its publishing efforts, American ceramists who wanted to learn from each other's work usually had to arrange to have their papers printed themselves, then sent as individual correspondence to colleagues. One such letter requested a paper from Society founder Edward Orton, Jr.

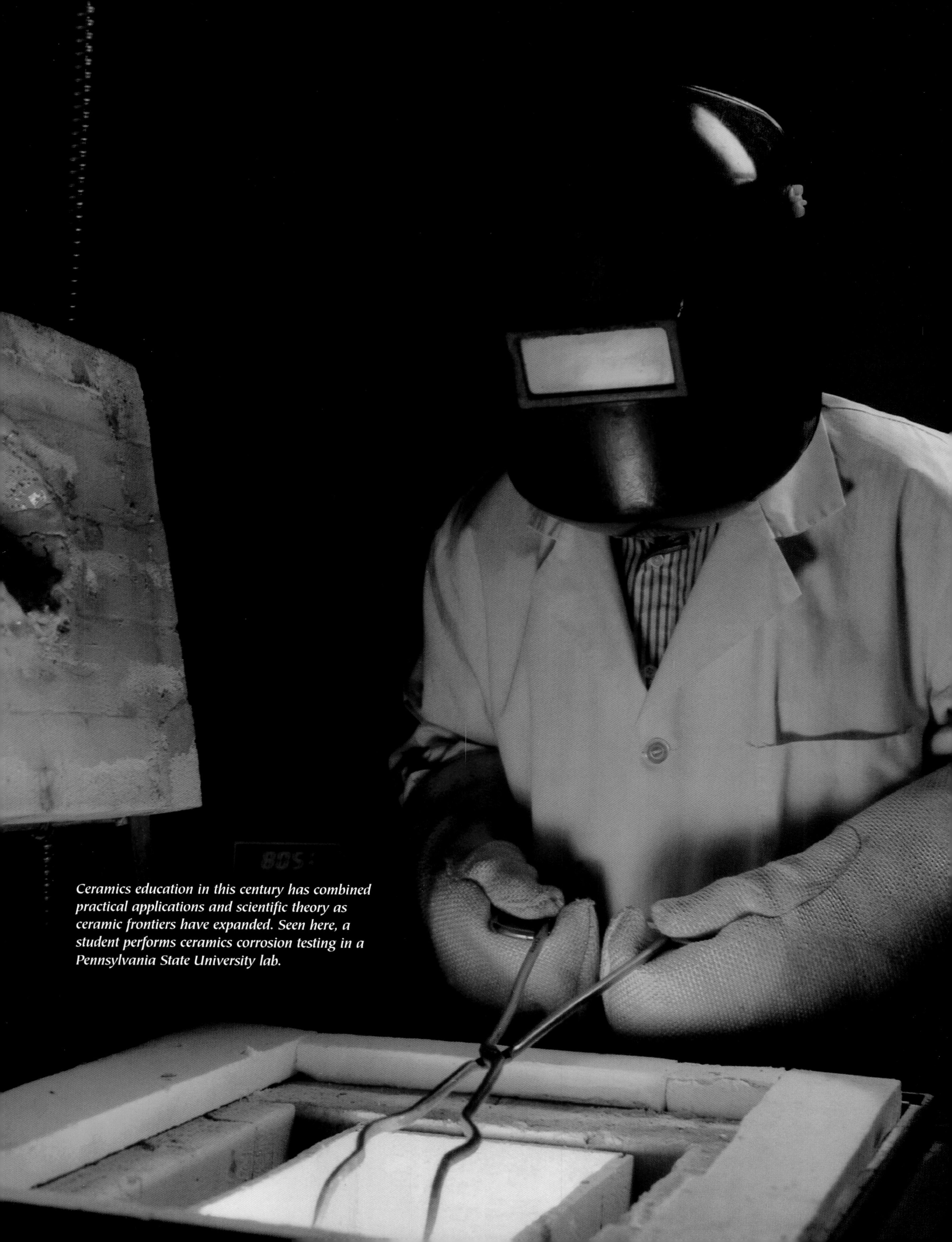

Ceramics education in this century has combined practical applications and scientific theory as ceramic frontiers have expanded. Seen here, a student performs ceramics corrosion testing in a Pennsylvania State University lab.

THE MAKING OF A CERAMIST

In May 1894 two rooms in the basement of The Ohio State University's College of Engineering were set aside to create a new department: Clay-Working and Ceramics. Founded with a $10,000 grant from the State of Ohio, the department was the first of its kind anywhere in the nation. • The course of study was the idea and personal project of Edward Orton, Jr., a young mining engineer. Orton had gathered his ceramic knowledge, piece by piece, from various jobs in coal mines, glass factories and steel plants. • In 1893 he was the superintendent for a factory that made paving bricks. Orton's factory, hit by an "acute crisis," closed, leaving him with some unexpected extra time. He used the opportunity to lobby for the passage of what some people dubbed the "Mud Pie Bill" — legislated funding to establish a school to train professionals in the growing industrial ceramic industry. • Four months later Edward Orton, Jr. taught his first class in ceramic engineering. Five years later he led the way to founding The American Ceramic Society — initially with two of his own students.

(Above) Born in Worcester, England, Charles Fergus Binns immigrated to the United States in 1897, eventually accepting the director's position at the newly formed New York State School of Clay-Working and Ceramics (now the New York State College of Ceramics at Alfred University). A charter member of The American Ceramic Society, a Society past president, and former General Secretary, Binns inspired ceramists and ceramic education.

(Above, left) Orton Hall at The Ohio State University housed not only the first formal ceramics program in the United States — established in 1894 with a $10,000 grant from the State of Ohio — but it also was the first "headquarters" of The American Ceramic Society.

More than a century after Orton's pioneering work, the industry and The American Ceramic Society are both influential in preparing young people for careers in ceramics. Today 11 schools are accredited in ceramic engineering in the United States. Enrollment has ebbed and flowed throughout the years, influenced by the growing prominence of science, events such as world wars and commercial trends such as contemporary applications for synthetic materials.

Charles Fergus Binns (center of photo) was the first director of the ceramics program at the New York State School of Clay-Working and Ceramics, the first program to accept women.

EARLY YEARS

Edward Orton, Jr. began the ceramics program at The Ohio State University (OSU) believing that ceramics lacked the structure needed for teaching new entrants in the field. Mining and metals had teaching programs, but most knowledge in the ceramic industry was gained only through practical experience, not through scholastic learning. It was a disadvantage that Orton wanted to overcome.

Orton's first program was only a two-year course of study, tucked quietly away within a larger program. Orton himself wrote that the early ceramics program was looked upon by the representatives of the academic courses at The Ohio State University as a fantastic perversion of the purposes of a university. Many sarcastic comparisons were suggested as to the need of courses in bread baking, making mud pies and similar humorous suggestions.

Out of the combination of personal need and a realization of the widespread need of others, the determination to start this school took form. It was a rash, foolhardy venture. It never could have taken place in any other country but our blessed America. But here, thank heaven, anybody can usually get the chance to make a fool of himself in his own way, at least once - and sometimes his folly leads to success.

Edward Orton, Jr., 1925

The jokes didn't keep Orton's dream from growing. The program for ceramics eventually metamorphosed into a four-year program that offered a degree: Engineer of Mines in Ceramics.

From 1894 to 1930 other universities gradually began to form their own ceramic engineering programs:

- New York State School of Clay-Working and Ceramics at Alfred University: 1900
- Rutgers University: 1902
- University of Illinois: 1905
- Iowa State College: 1906
- University of Washington: 1919
- West Virginia University: 1921
- North Carolina State University: 1923
- Georgia School of Technology: 1924
- Missouri School of Mines (now University of Missouri – Rolla): 1926

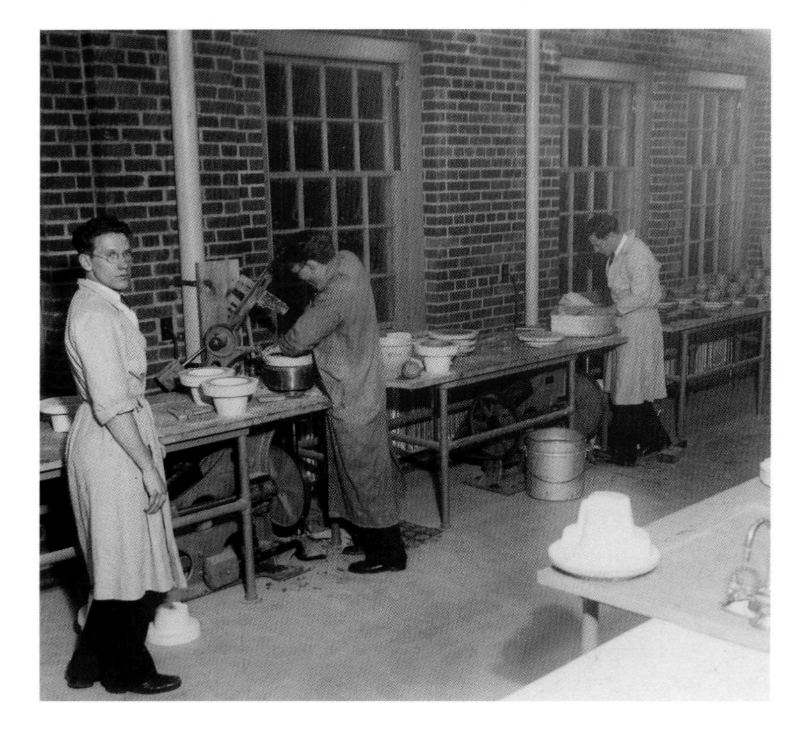

Students worked in the plaster shop at the New York State School of Clay-Working and Ceramics, formed in 1900, now the New York State College of Ceramics at Alfred University.

Charles W. Rolfe, a geology professor at the University of Illinois, formulated a state plan for ceramic education. Mainly because of his efforts, the Illinois legislature passed a bill in 1905 providing for instruction in ceramic technology at the University of Illinois. Rolfe was appointed director of the program, and Ross C. Purdy was appointed Instructor.

Clay-working methods in the 19th century, at factories in the United States such as this one, were passed along more by experience and apprenticeship than by scientific theory and formulas.

- University of Alabama: 1928
- Pennsylvania State College: 1923
- Massachusetts Institute of Technology: 1930

The first ceramics schools were formed with Orton's purpose in mind: to institutionalize the industry by educating aspiring ceramists with academic and scientific methods. At first the curriculum's importance was questioned in both the ceramic industry and the engineering colleges. Charles F. Binns from Alfred University addressed skeptics in the 1902 edition of *The Clay-Worker:* "Engineering, metallurgy, mining and a thousand other arts have taken advantage of higher education in the last twenty years. The clay worker has more need than them all and must not be denied."

The industry was rapidly expanding at the time, from small operations that made a limited number and variety of products to massive factories offering their customers myriad products sold by teams of salesmen. The time had come, Binns said, for education to step in and take the industry to the next level.

The first publication to answer the challenge was the scholarly *Journal of the American Ceramic Society*, created in 1918. Describing itself as "a monthly journal devoted to the arts and sciences related to the silicate industries," the *Journal* published original papers by members of the ceramic community. In keeping with the scholarly tradition rapidly being established in ceramic education the *Journal*'s first issue warned:

> In the consideration of papers offered for presentation or publication those papers containing matter readily found elsewhere, those specially advocating personal interest, those carelessly prepared or controverting established facts, and those purely speculative or foreign to the purposes of the Society, shall be rejected.

Today a variety of Society publications play an important role in ceramic education.

WORLD WAR I

World War I began on horseback and ended on wings. It sparked technological innovations from airplanes to can openers and brought the wave of the Industrial Revolution to a crest. The changes raised interest in ceramic education. Industrial research became a necessary fact of life for companies and a growing concern for universities where pure research had been the norm.

The lead editorial of the February 1920 edition of the *Journal* remarked:

> Before the war, research was looked upon as an interesting side issue of academic life, very indirectly related to factory management, but from the financial standpoint to be considered a non-essential luxury. To-day industrial research is not only respected; it has reached the enviable, but precarious, position of being a fashionable investment.

Orton's original curriculum — revolutionary in 1894 — was still being applied in 1924, 30 years after the founding of the OSU ceramic program. Some of the Society's members initiated a meeting for ceramic manufacturers and scholars on the 30th anniversary of the school's founding to update the curriculum, improve research and evaluate teaching methods.

Industrial ceramics courses began to spread to high school curricula. In 1925, the *Bulletin* announced the first high school courses in ceramics, to be offered in East Liverpool, Ohio, with funding from ceramic manufacturers in the town. A year later, a *Bulletin* update on the status of high school ceramic education reported:

> The Lincoln High School in Los Angeles has an $80,000 ceramic building under construction to care for its rapidly growing ceramic department. . . .There are several high schools, like Schenley in Pittsburgh and Central in Wheeling, which have highly developed ceramic art and ceramic science, technology and art.

The article advocated Society members getting involved in this growing trend by encouraging teachers to enroll in post-graduate ceramic programs at colleges and universities and instructing teachers to create programs that incorporated the artistic and scientific aspects of ceramics.

Ceramic education continued to spread on both a national and international level. In 1930 Orton wrote a paper for The Ceramic Society in Great Britain, outlining the changes during the 36 years since he had formed his program at OSU. He wrote that 16 colleges and universities in the United States and Canada offered four-year courses in ceramic technology; from these at least 750 men had graduated since 1900. He also remarked on the phenomenal growth in student enrollment, which by 1930 totaled 750 — the same number that had graduated in the previous 30 years combined.

Funding for ceramic education programs also had rapidly increased. Orton estimated that it cost no less than $250,000 to buy the equipment needed for a ceramic program and that universities would spend $850,000 to $1 million on buildings for ceramic education.

Students at the New York State School of Clay-Working and Ceramics (now the New York State College of Ceramics at Alfred University) load brick into a kiln in the early 1900s. The firebrick at the center of the photo reads "State School of Ceramics—Alfred, NY."

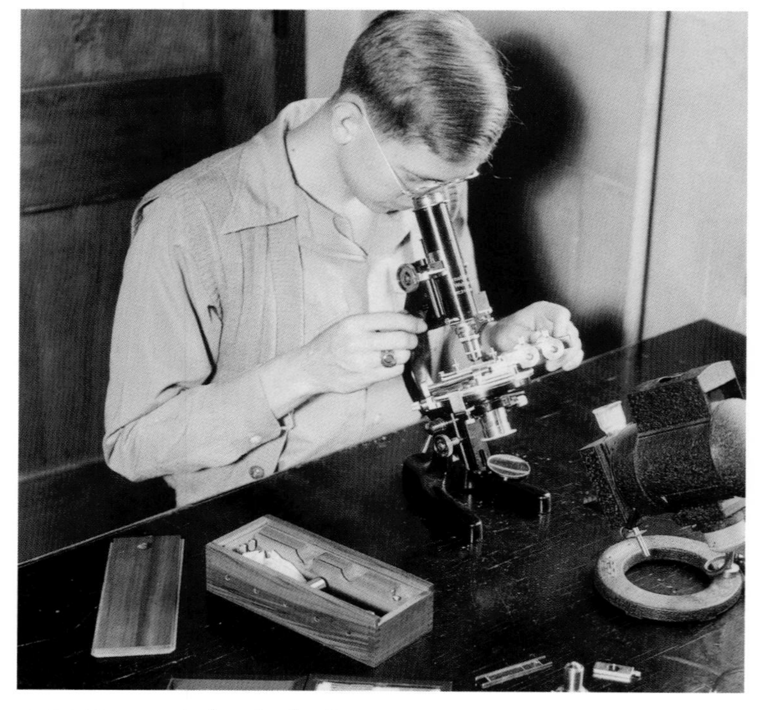

A 1940s-era student in the Pennsylvania State University ceramic engineering program measured a refractive index.

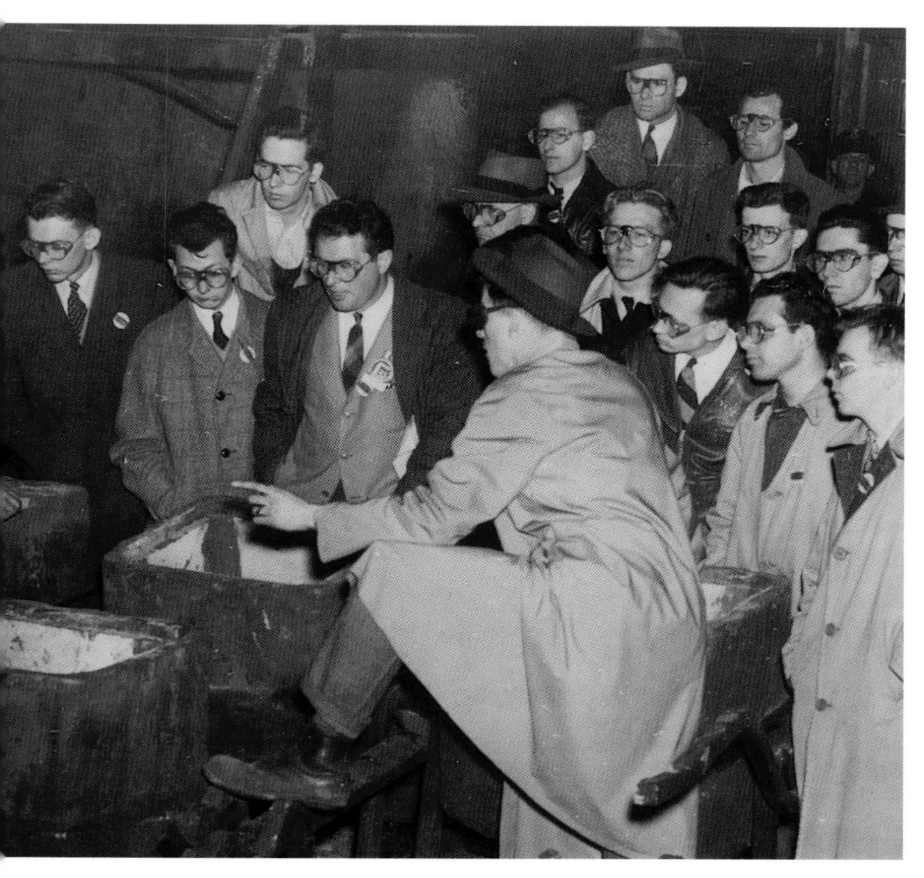

Ceramic engineering students clad in safety goggles visited a plant on one of the week-long industrial plant tours, a highlight of the senior year, toward which they had saved $50 a semester throughout their undergraduate days.

In 1952, two students examined material of a ceramic crucible with Dean John F. McMahon (center, seated), a past president of The American Ceramic Society, and glass professor Van Derck Frechette, both of the New York State College of Ceramics at Alfred University.

It was, he admitted, amazing growth for an idea that started out with $10,000 in grant money and two rooms in a basement.

THE 1930S

In 1934 the curricula for undergraduates studying ceramic engineering, particularly at OSU's School of Engineering, came under Society scrutiny. When the school's dean made curricular changes the Society saw as detrimental — increases in general education or cultural classes and reductions in engineering-related classes such as mathematics and chemistry — the Society published an editorial in the *Bulletin* outlining its qualifications for reviewing and recommending ceramic education curricula:

> The American Ceramic Society is exclusively devoted to the promotion of the ceramic arts and science in the interest of ceramic industries, and thus it is within the scope of the Society to consider what preparation the ceramic school graduates should receive for engineering and technical service in the ceramic industries.

One solution to the problem was offered two years later, when the Society's Committee on Ceramic Education discussed and eventually advocated accreditation for ceramic education programs in engineering schools throughout the country.

In addition to the Society's fight to influence the nation's ceramic education curricula, it continued to battle misconceptions about ceramics and its importance. Ceramic education had never been officially recognized by engineering educational organizations as a legitimate area of study, something the Society hoped to change.

A.F. Greaves-Walker, chairman of the Committee on Education, presented a paper at the 1937 Annual Meeting of the Society for Promotion of Engineering Education. During his presentation he "attempted to outline the history and give a definition of ceramic engineering and the scope of its field." He apparently was successful. Before the meeting had concluded the group voted to organize a new class of the Society called the Institute of Ceramic Engineers (later to become the National Institute of Ceramic Engineers) whose goal it was to advance the science and practice of ceramic engineering in the public interest. It was another step toward full recognition of ceramics in engineering circles.

The next year, the Association of Ceramic Educators successfully petitioned to become a second class of the

Society known as the Ceramic Educational Council for the purpose of stimulating, promoting and improving ceramic education.

WORLD WAR II

World War II threw ceramic education and the entire nation into upheaval. A 1942 edition of the *Bulletin* discussed wartime changes for the ceramic community. One advertisement in that issue, optimistically predicting the end of the war (in reality about three years away), read like a recruitment poster: "Postwar Security is Essential — There Must Be No Slacking in Ceramic Research and Education. THE AMERICAN CERAMIC SOCIETY Is Your Opportunity."

The ceramic industry, like so many other industries of the time, experienced a severe manpower shortage, particularly of trained technical workers. The number of students studying ceramics at the time also had dropped to critical levels. The Society's Ceramic Educational Council conducted a survey of ceramic education programs and found that nationally only 71 graduating seniors would be available for jobs in the ceramic industry once they left college; the rest went directly into military service.

To offset this trend, the Council advocated deferring military service for ceramic students, which would enable them to move directly into the industry upon completing their training. In a notice published in the *Bulletin*, the Council listed graduation dates for several colleges and universities that offered ceramic education programs. The notice warned, "It will be obvious from the foregoing evidence that companies desiring ceramic graduates should take immediate steps to contact them. The number of men available is extremely limited."

In addition to these efforts, steps were taken to get high school-age students involved in ceramic education and production. A 1942 *Bulletin* article entitled "Ceramics in NYA: A War Emergency," showed the efforts of the National Youth Administration for Ohio, which ran ceramic workshops that aspired "to train youth for various ceramic industries."

The article showed signs that the ceramic industry, like most others in the nation at that time, was opening its doors to women. It was even written by a woman, Josephine Gitter, who said somewhat tongue-in-cheek: "The original crew of ten boys grew, girls were added. . . Fewer girls go into private employment than boys, but it should be remembered that it is only within the last few months that industry has found it increasingly necessary to employ girls to replace boys."

In 1952, the members of Pennsylvania State's Keramos chapter posed for this photo. Keramos, ceramic engineering's honor fraternity, has chapters in ceramics schools across the country. The fraternity aims to promote academic excellence in ceramic engineering and increase community awareness of ceramic engineering.

In 1958, ceramic educators held their second conference. Since forming in the 1920s, the Committee on Ceramic Education has addressed issues such as curricula components, nontechnical courses and enrollment levels in ceramic schools. As ceramics have evolved, educators have worked to maintain high standards, ensuring that consistently high-quality graduates are available to the industry.

Lab equipment of the late 1940s, similar to the setup this student was using, became increasingly technical.

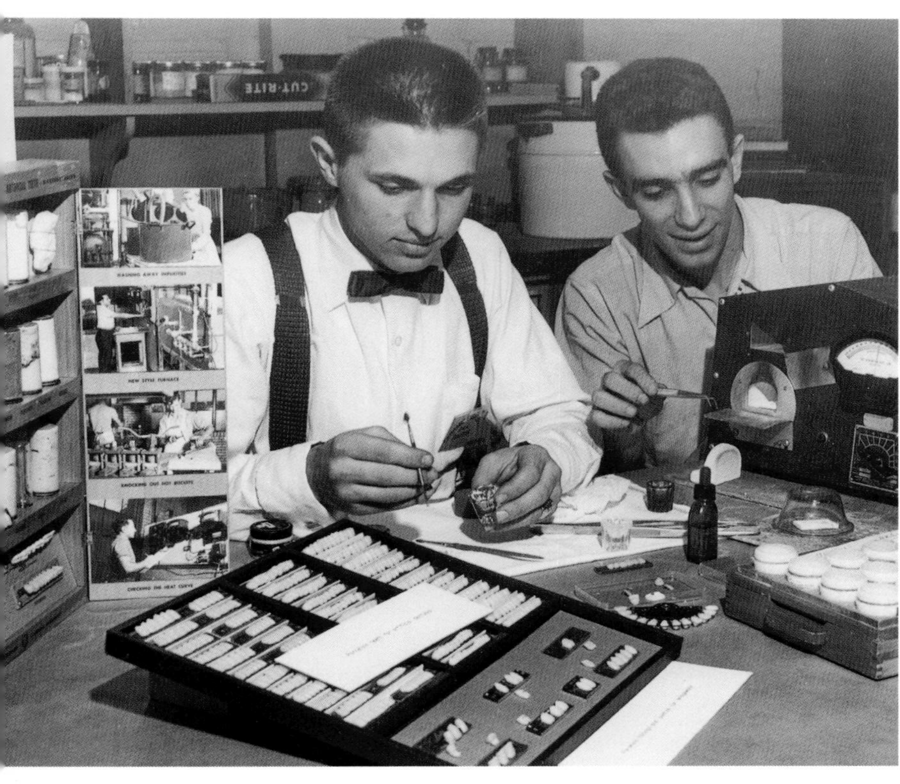

Two students work on a dental ceramics display being prepared for the 1956 Open House at the University of Washington.

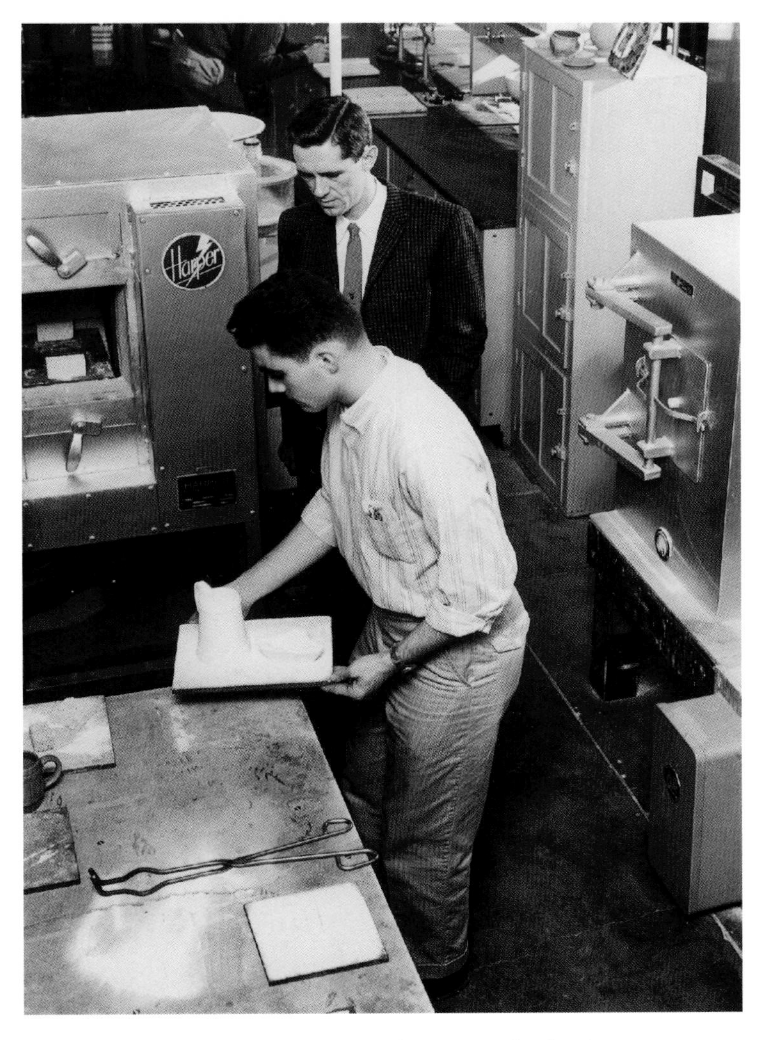

Students learned to load a kiln with items to be fired at Pennsylvania State University in 1959.

A photo accompanying that article showed a young woman being trained to work with a ceramic mold that created dinnerware to be used by soldiers. In the war years and after, more women worked in factories with industrial ceramics, but an increasing number also enrolled in university ceramic programs.

POST-WAR READJUSTMENT

When World War II ended, a period of change overtook the ceramic education community. After years of severely reduced enrollment in ceramics courses at the university level, students were back in school. In 1948 colleges and universities were packed with young men back from the war, a large number of them pursuing degrees under the GI Bill. Many of them headed for engineering and technical areas of study, and many schools took advantage of increased enrollment, making positive changes in their curricula and courses.

In 1948 demand also was extremely high for industrial ceramics instructors; enrollment of upperclassmen in ceramics courses had more than quadrupled since a 1942 survey. In the United States and Canada, 900 undergraduates were studying ceramics at engineering schools.

Ceramic education officials within the Society were so optimistic about future enrollment numbers that they began selecting only the best and brightest students for ceramics programs, a far cry from the severe shortage of only a few years before.

THE 1950S

The postwar boom in ceramic enrollment was short-lived. As Society President R.R. Danielson wrote in 1954, "It was only a few years ago that the universities, including the engineering schools, were swamped with students. It did not seem possible that four or five years hence the engineering and scientific schools would be aggressively seeking students."

The number of ceramics graduates peaked in 1950 at 330 students. As GI Bill students earned degrees and left universities, no new influx of students appeared to fill the gaps. The following years brought fewer and fewer graduates and dismal predictions from educators about future numbers. Engineering schools in general suffered from the problem, but ceramics — a lesser-known area of study — felt the decline more acutely. Two Canadian programs shut down entirely because of the student shortage.

In 1952 W.W. Kriegel, president of the Ceramic Educational Council, wrote in the *Bulletin*:

> Many causes might be cited for the low enrollment in ceramic engineering curricula. However, it is the

> common belief of educators that the principal cause is the abysmal ignorance of the high-school students and the public in general about the field of ceramics. Many do not know the word. Those who have heard of it think only in terms of museum and hobby pottery.

It was time for some active recruiting. The Ceramic Educational Council created a public relations program designed to increase public awareness of — and perhaps enrollment in — ceramic engineering. The plan recommended distributing a brochure on ceramic opportunities to prospective students, a film that could be shown in schools and on television, and a continuing scholarship program with special emphasis placed on promoting the opportunities ceramics afforded students.

David Kingery (left) listened to a student make his point in 1965 at the Massachusetts Institute of Technology's Center for Materials Science and Technology. Kingery's texts on ceramics are standards in ceramic education.

One recommendation proposed changing every book in the United States that described ceramics. It advocated "that The American Ceramic Society place its prestige behind the full definition of ceramics, making every effort to correct erroneous information in encyclopedias, dictionaries, high-school texts, etc."

But as the decade progressed, some ceramic education experts began to realize that conventional public relations efforts only went so far. Robert Twells, Society president in 1955, recalled how his career began, and it had nothing to do with brochures or films.

Twells wrote that during a Christmas vacation in 1914 an engineering student named Frank Hunt visited the Key-James Brick Company in Chattanooga, Tennessee. At the time Twells was a young foreman working in the shipping department. Twells and Hunt had a long conversation about ceramic engineering, one that convinced Twells to move to the University of Illinois and earn his degree in ceramic engineering. "By such a slender thread our destiny sometimes hangs," Twells wrote.

> Now, I might eventually have gone to college and might have taken ceramic engineering, because I had long wished to become a ceramic engineer. But the prospect seemed then about as remote and impossible as a trip to the moon. Certainly the contact with Frank Hunt crystallized my ideas and made me feel that the dream could become a reality.

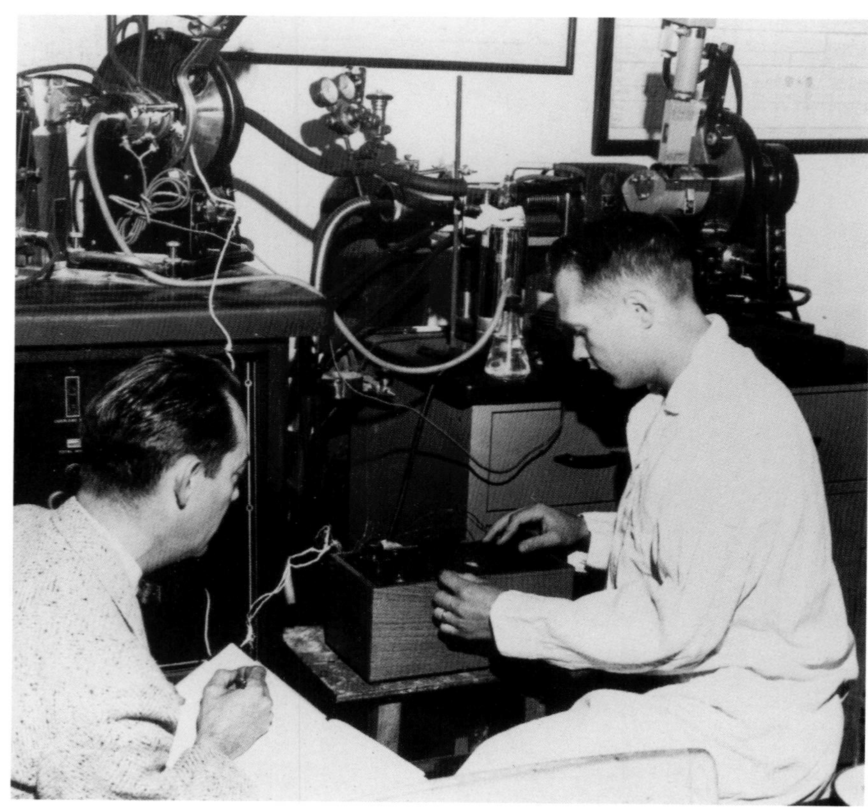

James Mueller (left), a past president of the Society, worked with a tudent on an X-ray diffractometer at the University of Washington in 1964. His educational legacy has continued in the Society — the research library at ACerS headquarters is named after him.

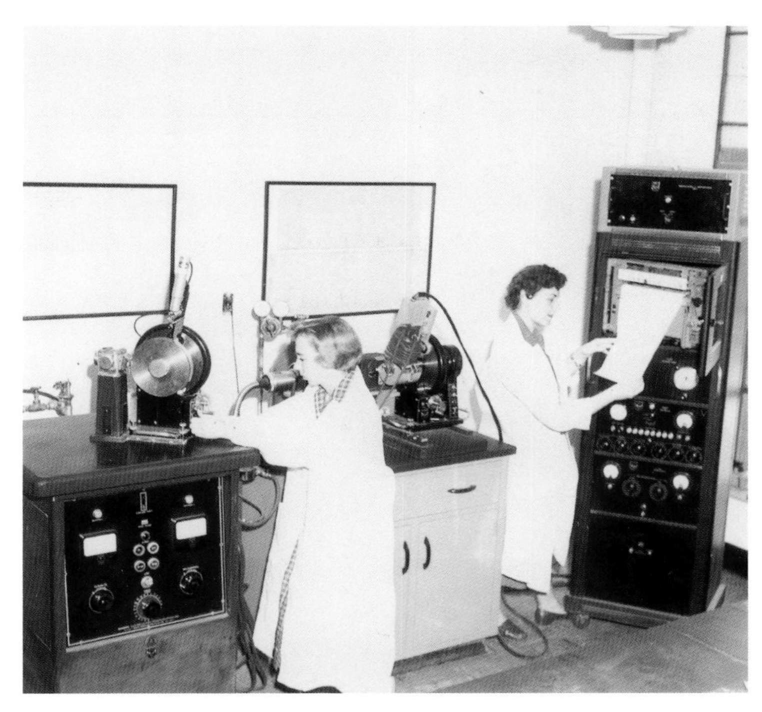

Students work in an X-ray lab at the University of Washington. By the 1950s, women working in ceramic engineering labs had become more numerous.

In the design kiln room at the New York State College of Ceramics at Alfred University, students in 1962 loaded a gas-fired kiln.

Such stories encouraged Society members to address high school and civic groups, publicity that Twells said "takes little or no money but is the most valuable kind of publicity which can be obtained." Several generations of ceramists bear out this experience. Most Society members trace their interest in ceramics to an inspiring ceramist.

The 1950s also brought a new trend to university-level ceramics — training students as managers, not just engineers. Companies needed recruits who could play the dual role of highly skilled engineers and team managers, but big corporations such as General Electric and DuPont were the only ones that could offer their employees such intensive instruction. Once again ceramists had to learn an important skill in the school of experience. Aware of the advantages for students who received management training in college, the Society began to develop a course that would teach young ceramic engineers the specifics of being a middle plant manager.

Ceramic engineering enrollment took a modest upturn in the late 1950s, due in part to the increased use of ceramic materials in technologically advanced electronics and nuclear power. The space program, developed in part as a response to Russia's 1957 *Sputnik* launch, increased the need for ceramics. A 1957 survey showed a 12 percent increase in ceramic engineering enrollment between 1950 and 1956.

THE 1960S

Although the new decade began with promising enrollment numbers, the chairman of the Society's Committee on Education in 1960, John H. Koenig, urged members to help increase enrollment by reaching out to students to teach them about career opportunities. "If ceramic graduates are desired, neither industry nor schools can afford a complacent attitude," he wrote.

Enrollment figures continued to be an important gauge of the state of ceramic engineering education. A 1962 *Bulletin* article written by James Mueller — past president of the Ceramic Educational Council — reported some encouraging news. Ceramic engineering enrollment, he wrote, increased (albeit a modest 1 or 2 percent), while engineering enrollment had decreased by 13 percent overall.

Other indicators of the health of ceramic education also concerned Society members, who were interested not only in how many students enrolled but what they were being taught and whether other contributions to ceramics were being made in university laboratories.

Mueller addressed the importance of examining the curriculum and promoting research in the field.

Curricula that trained students specifically for careers in ceramics were eroding, Mueller wrote. And two noticeable trends had developed in research: Project money was increasingly coming from sources outside universities, and the research itself — once conducted for its own sake — now had a more defined purpose. He stressed the importance, particularly because of the space program, of encouraging students to seek ceramic engineering degrees and significant research opportunities. In 1962 Mueller anticipated how expanding technologies would change the future of ceramic education:

> Students that enter our departments next fall will be graduating at a time when our country will have already had several men in orbit in a single capsule . . . We will probably have landed robot equipment on the moon. We may have space stations in orbit around the earth. Your own minds can certainly visualize others of today's scientific developments that will be considered every-day occurrences by then.

The following decades saw his predictions come to pass.

Organizations that promoted ceramic education also experienced changes of their own. In the 1960s, the Ceramic Art Educators separated themselves from the Ceramic Educational Council and The American Ceramic Society by forming the National Council for Education for Ceramic Arts (NCECA). This separation left the Ceramic Educational Council composed mostly of ceramic science and engineering educators. Renewed joint publishing and programming efforts in the last few years between NCECA and the Society have revitalized the Society's commitment to education in the ceramic arts.

THE 1970S AND 1980S

During these decades ceramic education enrollment increased overall, periodically bouncing up and down from year to year. In the 1976-1977 academic year ceramic engineering undergraduates across the country earned 161 bachelor's degrees. By 1980 that number was 262. Five years later it had risen to 310, where it remained relatively steady for the rest of the decade.

The demand for ceramic engineers at companies around the world created faculty shortages at engineering schools; there simply were too few professors to teach the number of students who wanted to enter

Students attended a Poster Session at a Society meeting in 1976. Student involvement in annual meetings is an important part of ceramic education.

Dean John F. McMahon congratulated a graduating student at Alfred University.

Engineering and science students at Alfred University ran pressed ceramic ware in the first-year engineering foundation course during the 1980s. Lovingly called the "mud lab," the course was developed to give students hands-on experience with ceramic products.

A student uses the secondary ion mass spectrometer at the Pennsylvania State Materials Characterization Lab. Increasingly sophisticated equipment has come to be part of ceramic education as processes and materials used in ceramic innovations have changed.

Once a rarity in professional ceramics studies, women today make up an increasing percentage of graduating engineers and scientists who will work in the ceramic industry in the 21st century.

the field. A 1982 article that discussed the cause of the shortage cited not only increased enrollment, but a 33 percent gap between salaries of privately employed engineers and those working as college or university instructors. Faculty was leaving academe for better pay and better equipment in the private sector.

In response to these problems, NICE has redoubled its efforts to support ceramic engineering programs and students. NICE initiated a Student Scholarship Program in 1985, distributing scholarship funds to those schools with ceramic engineering programs accredited by the Accreditation Board for Engineering and Technology. The Student Congress was initiated in 1989 through the efforts of John D. Buckley and David E. Clark for the purpose of grooming student leaders for the ceramics profession.

Women in ceramic engineering programs continued to be in the minority throughout the 1970s and 1980s (usually 20 to 30 percent of total enrollment in the 1980s), although the environment had clearly improved since a 1971 survey in which Alfred University listed the placement of its 67 recent graduates: "27 percent grad school, 10 percent military, 33 percent industry, 27 percent unemployed but some interviewing, 3 percent housewives."

One of the most important changes in ceramic engineering education came with the advent of continuing education through the use of "short courses." For the first time in September 1975, a short course on "Kinetics in Ceramic Processes" was offered in Bedford, Pennsylvania. It would be the first of many short courses in what is now a NICE program of short courses that are an integral part of the Society's Annual Meeting technical program. By 1988 the Society had acquired the Ceramic Correspondence Institute which, under the guidance of the National Institute of Ceramic Engineers, offers continuing education programs, seminars and correspondence courses in 29 different subjects.

THE 1990S

College and university ceramic engineering enrollment, still a favored gauge of overall interest in ceramics, began a slow but steady decline in the 1990s. Coinciding with this decrease, a growing number of individuals now believe that ceramics is not a separate area of study, but a part of materials science.

Dennis Readey, a Society past president and a professor of ceramic engineering with the Colorado School of Mines, sees the job market for ceramic engineers a bit differently. He believes that beyond traditional ceramic industries, such as dinnerware and brickmaking, not as many employers are hiring graduates for their ceramic engineering degrees. He was formerly a professor with The Ohio State University ceramic engineering program, where undergraduate students still can pursue a degree in ceramic engineering through the Materials Science and Engineering Department. He has come to believe that "undergraduates in ceramic engineering are more limited in their careers than they probably should be. This is because the industrial community really doesn't understand ceramics, and they hire materials science graduates instead."

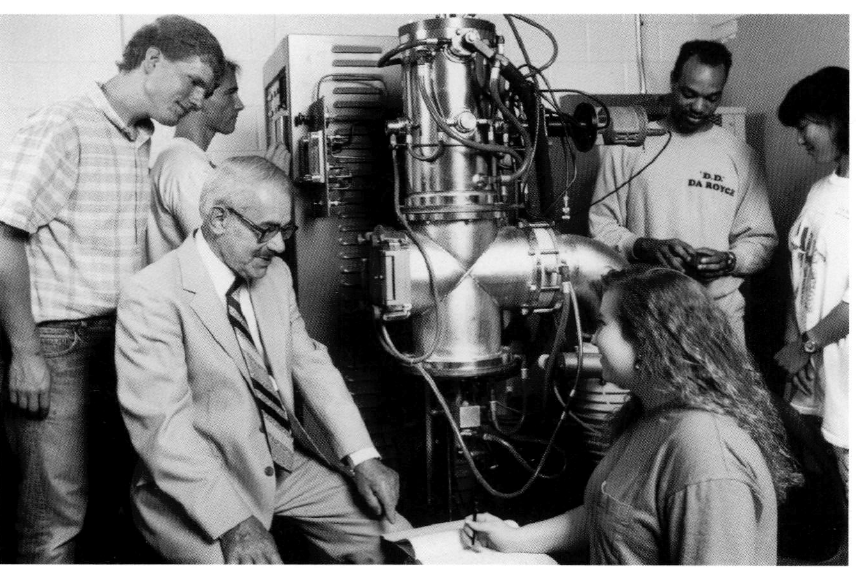

Robert J. Campbell and students worked with a high-temperature Vacuum Furnace at the University of Washington in 1991.

Some educators lament this change. Delbert Day, a Society past president and a professor of ceramic engineering at the University of Missouri-Rolla, said he supports maintaining ceramic engineering as its own area. "I'm an optimist, so I suppose my glasses are a little rose-colored," Day said. "But I think the state of ceramic engineering is good. I will always feel there is a place for ceramic engineering."

Ceramic education and research continue to be strong in many institutions, particularly because of the new applications that are opening up in fields such as telecommunications, computing, medicine and dentistry. Day said he finds that employers still want to hire engineers with degrees in ceramic engineering for their specialized knowledge.

Some programs compromise with combinations. Denis Brosnan, a professor at Clemson University's Ceramic and Materials Engineering Department, said that in his program, although the word "materials" is in the department's name, ceramic engineering is the ongoing focus of the program. Other programs in ceramics have been absorbed into materials science departments to accommodate the changes within ceramic education that resulted from increases in the variety of synthetic materials available to engineers today and substances that produce similar results and use similar processes as ceramic materials.

Like so many disciplines on the eve of the millennium, ceramic education is changing. The New York State College of Ceramics at Alfred University, for example, experienced changes similar to many other programs. Alfred University's program once offered only one degree, in ceramics. It now offers a bachelor's degree in fine arts for Ceramic Art as well as bachelor's degrees in Ceramic Engineering, Glass Engineering and Materials Science & Engineering.

(Center) Future ceramic engineers work with Westerville, Ohio, fourth graders at Society headquarters in a student outreach program. (Bottom) Pennsylvania State's 1953 Ceramic Science Open House included demonstrations of various ceramic science principles and techniques for elementary-school children in the area.

(Above) Student field trips to the Society's Annual Meeting Exposition provide opportunities for ceramic engineers to meet the ceramics professionals and customers of the future.

(Below) University of Illinois students attend 1997's annual Engineering Open House, where a ceramics student demonstrates the basics of ceramic engineering. Educating prospective ceramics students continues to be important to the future of the industry.

THE FUTURE FOR CERAMIC EDUCATION

As ceramic engineering evolves, research in a number of areas could bring new opportunities for ceramic engineers.

Exciting research within ceramic education also could make substantial contributions to several non-ceramic fields and encourage training new ceramists. Advances in electronics, telecommunications and medicine are creating future possibilities for ceramic engineering.

Readey says he thinks students will continue to be interested in ceramic engineering despite the trend toward materials science programs. "Particularly at the graduate level, a lot of students view ceramic materials as a growing, wide field of application. They see it as a field with a lot of opportunity."

The ceramic industry and the Society will keep working to improve the public understanding and perception of ceramic engineering. Dennis Readey said that he and other ceramics educators and professionals would like to see an open, ongoing discussion of strategies for increasing public knowledge of ceramics.

Despite the difficulties ceramic engineering education programs face, educators say they will continue working to ensure that students pursue degrees in the field. They are committed to passing on the exciting potential of ceramics that Edward Orton, Jr. instilled in 1894 in his basement classrooms. ▲

(Left) Students who attend Society annual meetings experience early the benefits of meeting regularly with colleagues.

dvances in technology have become a key driver of today's highly competitive world economy. Ceramics play many essential roles in modern technology and are likely to be even more important as technology continues to advance. Needs and opportunities for ceramics in the future are particularly evident in such areas as computing, communications, defense, medicine, pollution abatement, and transportation. A look at trends in technology in general and the associated development of ceramic technology provides a perspective for thinking about the future. • Progress in commercial technology sometimes results from the accumulated effect of many small advances. These innovations sometimes had their effect in the direct use of ceramics and in other cases indirectly through the use of ceramics in processing or as components in products. Worldwide conditions in technology generally parallel those in the United States, so the pattern of developments in the United States provides a useful and broad perspective. • The status of ceramic science and technology a century ago was that of art and craft in the process of becoming a field of science and technology.

(Above) For the growing semiconductor industry, vertical wafer boats are important in vertical furnace applications.

(Opposite) These ceramic abrasive grains are used for industrial pressure blasting. Their uses include cleaning metal and machine parts, as well as etching glass and stone.

ADVANCES IN CERAMICS

by

John B. Wachtman, Jr.

CERAMIC TECHNOLOGY BEFORE 1899

Ceramics have been made and used since before the beginning of written history. Innovations in ceramics have come from many sources. Ceramic engineers have led the changes of the last century, but contributions have also come from other disciplines including chemistry, physics, civil engineering, mechanical engineering, geology, metallurgy, chemical engineering, and materials science.

A large body of practical art existed in 1899 and was the basis for much of the manufacturing of ceramics at that time. By 1899 the beginnings of ceramic science also existed. In Germany H. Seger in 1874 contributed the "rational analysis" of ceramic raw materials into clay, quartz and feldspar with various accompanying rules and also developed his pyrometric cones for better firing control. In the United States, Josiah Willard Gibbs enunciated the phase rule in 1878.

Worldwide, science and ceramic technology together were making important advances. The traditional craft of pottery and tableware making was influenced by better bodies and glazes as well as better understanding of the effect of firing conditions.

When The American Ceramics Society was founded, making a tea cup was not much different than pictured in an 1878 illustration, "The Handler." During the Society's early years, the ceramics industry was truly clay-based and the most common ceramic products — bricks, sewer pipe, tiles, glass, dinnerware and china and fine art ceramics — were made using techniques that had, in most instances, been around for centuries. Brickyard and pottery operators jealously guarded their individual formulas and processes. The American Ceramic Society played a large role in turning the industry from narrow commercial interests to a broader scientific outlook.

The electrical and chemical industries required porcelain for containers and liners and electrical insulators. Bell-shaped insulators for telegraph poles were in mass use by the 1850s. The automobile developed in Germany beginning in 1885 when Karl Benz built his first motor vehicle using an Otto-type engine with a spark plug dependent on a ceramic insulator. Many chemical processes were carried out in ceramic containers or containers lined with ceramics.

Cities grew as innovations in ceramics made possible more and better building materials. Brick and tile for general construction have been used since prehistoric times, but many types now common have come into mass use since about 1850. Machinery for the mixing of ceramic raw materials and forming of unfired ceramics was originally worked manually. In England James Watt's steam engine was developed for steam-driven mills, breakers and

A steam shovel was used to remove overburden from clay in 1900, with the clay itself dug by hand. Machines promised opportunity to factories and potteries. When the Society was founded in 1898, electricity was rare, but as the electrical grid spread across the landscape, the ceramic industry eagerly employed new powered machinery.

The historical product of the nation's oldest pottery maker, The Pfaltzgraff Co, which made salt-glazed pottery beginning in 1811 and became a sophisticated manufacturer in this century. By 1900, changes were under way that would streamline all traditional ceramic industries, making them more efficient and their products of consistent quality with greater affordability. Hand shaping of clay became obsolete except for the fine arts potter. New forming techniques included slip casting, extrusion, uniaxial pressing and, in the 1920s and 1930s, isostatic pressing, tape casting and injection molding.

mixers for ceramic raw materials. Many advances in machinery for ceramic production, including presses and auger machines were introduced from 1850-1870. As electric motors and internal combustion engines became available late in the 19th century they replaced steam in powering ceramic processing equipment.

The Romans found that a mixture of lime and volcanic ashes would develop strength under water, but today's cement industry really goes back to the studies of John Smeaton in 1776 in connection with the building of the Eddystone lighthouse, when he found that by firing soft, clayey limestones he obtained a cement that would harden underwater. The firing converted calcium carbonate at least partly to calcium oxide that would hydrate under water and thus harden. A wide range of compositions of hydraulic cements is possible, and a series of investigators added other raw materials. The invention of portland cement (distinguished by containing silica and alumina in addition to calcium oxide) is usually credited to Joseph Aspdin in England in 1824. Reinforced concrete was first made in 1868; the rotary kiln for cement making was invented in 1873; and the manufacture of portland cement in the United States began in 1875. The long process of moving ceramics from a tradition-based craft to a science-based technology was under way in the 1800s, was accelerated by technical discovery and has continued unabated to the present day.

The United States imported its scientists and engineers in its early years. Although the United States continues to this day to import many technically trained people, a very strong domestic network of universities and technical societies was gradually built up. The development of science and engineering in the 19th century was accompanied by the founding of specific departments in universities and technical societies dedicated to newly emerging fields of science and technology. The Department of Ceramics started at The Ohio State University in 1894. Also important to the field of ceramics was the founding of The Bureau of Standards in 1901. This agency of the U.S. government subsequently became the National Bureau of Standards and later the National Institute of Standards and Technology as its responsibilities were broadened.

CERAMIC TECHNOLOGY SINCE 1899

Among the 20th-century developments that stimulated progress in ceramics were advances in science and technology in general, the rise of new industries, calamitous advances in military technology, and concerns for health, safety and the environment.

It is useful to think of ceramics in the 19th and 20th centuries in terms of three major effects. First, the rise of many new technologies outside ceramics called for materials with new and improved properties. This stimulated development of new ceramics for some of these needs. Second, advances in techniques for characterizing materials assisted in the development of totally new ceramics as well as leading to improvements in the properties of existing types of ceramics. Third, advances in mechanical and electrical engineering have been applied to existing processes of ceramic production. Innovation often involves a combination of these effects.

Advances in science and technology outside ceramics have had great effect on ceramics in the past 100 years. Events near the turn of the century set the stage for a still-continuing period of unprecedented growth in science and technology, with each assisting the other in a synergistic relationship.

W.B. Stephen of the Pisgah Forest Pottery about 1930 shaped a pot. While there was still room for craft, the rapid motorization of the industry would have stunned a 19th-century potter or brick maker.

Maxwell enunciated his theory of electricity in 1865 and Hertz in 1888 proved the reality of the electromagnetic waves that this theory predicted. G.J. Stoney in 1874 identified the discrete unit of electricity and later named it the electron. J.J. Thompson showed that cathode rays consisted of streams of particles and so established the reality of the electron in 1895, putting in place the building block upon which electronics is founded. Milestones in the subsequent development of radio communication include Marconi's transmission of radio messages across the Atlantic in 1901 and De Forest's development of the triode vacuum tube and its use in oscillators and amplifiers by 1910. In another stream of scientific development, Max Planck inaugurated the quantum theory in 1900 with his discovery of the quantum of energy.

Salt-glazed crocks and jugs were among the first ceramic products made in America.

His successors erected a theoretical framework that would provide understanding of electronic devices and quantitative modeling of them as well as suggesting new devices. It was the beginning of the age of electronics, an era that would build upon the stil growing age of electricity. The electronics and communications industries that grew with 20th century advances in science, especially physics, placed new demands on materials and greatly stimulated innovations in ceramics. The development of the transistor in 1947 and the growth of the computer industry from the 1940s to the present day have spawned innumerable spin-off and satellite industries with special requirements for ceramics.

The demand for whitewares and the ready availability of the necessary materials created a thriving ceramics industry east of the Mississippi in 19th-century America.

An airfloated ball clay bagging operation in the late 1930s supported the need for ceramic materials.

FERRO

IN CELEBRATION OF CERAMICS

We at Ferro Corporation celebrate with The American Ceramic Society on its 100-year anniversary and salute its many accomplishments.

For nearly 80 of those years, Ferro has been proud to contribute to the creativity and achievements in the ceramic industry and to serve as a high-performance partner to ceramic manufacturers worldwide. Ferro is the world's largest supplier of ceramic glaze coatings and a leading producer of pigments and colors for those coatings. Indeed, glass coatings have been a staple product for Ferro since our founding in 1919. We are also an important producer of specialty ceramics.

Our advances in ceramic glaze coatings and colors reflect our intensive research and development efforts, and especially our focus on creating value-added products and services which give our customers a market advantage.

Early on, we were first to make the production of glass coatings scientific. More recently, our lead-free glaze systems for bone and fine china offer our customers compliance with stringent environmental regulations as well as outstanding durability. Our abrasion-resistant ceramic glazes and dichromatic glazes have given customers new options for enhanced performance and appearance. And our centralized ceramic design and development labs provide tile manufacturers with exclusive designs as well as high-quality materials and manufacturing assistance.

Our focus on providing customer solutions, combined with strong R&D capabilities, has helped make Ferro a $1.4 billion global leader in a variety of performance materials, including specialty plastics and chemicals, to serve manufacturers around the world.

In the last 100 years (and particularly the last 50 or 60) a diverse field of high-technology ceramics — electronics, magnetics, optical, sensors, computer-related, etc. — based on synthetic raw materials has grown up parallel to the clay-based ceramic industry — whitewares, refractories, structural clay products, glass — that is still based largely on clay and other raw materials such as silica.

A development of great importance to traditional ceramics and to the high-technology ceramics of the last half century is the working out of ceramic phase equilibria. Of equal importance is the increasingly systematic compilation and use of this information in the discovery and processing of improved ceramics. Complete melting is important in the preparation of glass and the growth of single crystals from the melt. Partial melting is used in liquid-phase sintering of ceramics. Melting must be avoided or at least limited in the use of refractories for steelmaking and other high-temperature applications. The visual expression of Gibbs' phase rule in phase diagrams graphically presents the melting behavior and much other information vital to advances in ceramic processing and properties.

The lead in developing techniques for high-temperature phase diagram determination was taken by N.L. Bowen of the Geophysical Laboratory of the Carnegie Institution as early as 1914. He and his group developed the quench method requiring platinum sealing techniques, special furnaces and good temperature controllers. The development by Braggs and others of X-ray diffraction as a tool for crystal structure determination made it the primary characterization tool in phase equilibria studies. Petrographic microscopy however, is still an important supplement. Using these techniques Bowen and others, including G. W. Morey and J.F. Schairer, carried out a program of phase

In 1898 the Owens bottle machine was developed and was in factory operation by 1904. An early Society program included a visit to a plant to see an Owens machine operate. One version of the machine could produce 2,500 bottles per hour. But the Owens would soon be superseded by other developments in glass production, such as the Corning ribbon machine (late 1920s) that could produce 1,800 glass bulbs per minute.

American presidents like to entertain on American tableware, like the presidential china made by Lenox for the Truman and Reagan White Houses. During World War I, because the German porcelain to which American consumers were accustomed was no longer available, domestic makers stepped in to meet the demand.

The war caused a sudden spurt of development in ceramics, with industry members working to ensure a steady flow of ceramic materials and products to the country's industries and military — everything from spark plugs with demanding combat applications to optical glass for gun sights to abrasives.

For years ceramic tiles, terra cotta trim and brick were staples of the builder's trade. Determined to match the quality of the English wares at the Philadelphia Centennial Exhibition in 1876, the Celadon Terra Cotta Company built this terra cotta house in 1892. New, alternative construction materials soon became available to architects and construction engineers, and were aggressively marketed by their makers. By the 1930s terra cotta virtually met its demise as a major ceramic industry.

diagram determination and founded a school that soon had members in many other places, including particularly important groups at Pennsylvania State University and at the National Bureau of Standards. The latter group became the home of the international compilation and dissemination of phase diagrams that is now carried out under the auspices of The American Ceramic Society.

The development of the chemistry, thermodynamics and kinetics of point defects in crystalline materials in the last 50 years is another major advance with many implications. This science underlies the understanding of matter transport in sintering and creep. Point defect science also has a major role in understanding electronic properties of insulators and semiconductors and so of most ceramics.

In addition to looking at the effects of totally new industries on ceramics, consideration should be given to the way in which advances in manufacturing technologies affected the production of traditional ceramics. The ceramic and glass industry at the turn of the century was greatly and favorably affected by the general development of mechanical devices and electric power. Metals and alloys are most commonly made by casting from the melt, although sintering of metal powders is also used. Containing the metal melt requires ceramic refractories. More than 50 percent of refractories are used in the iron and steel industry. Changes in the demand for types of steel and advances in iron and steel making technology have caused great changes in refractories in the last 50 years. Clay refractories based on alumina-silica fireclays were traditionally used. In blast furnaces for ironmaking and open hearths for steelmaking, the traditional refractory lined walls have been largely replaced above the melt by water-cooled metal walls.

The Great Depression was responsible for a vast weeding out of the ceramic industry. Generally speaking, companies that had modernized and developed controls with trained engineers in charge were the least affected. Many antiquated plants were forced out of business. Some, like Franklin Tile shown in 1923 in its first plant, in a renovated apple butter factory, prospered. Franklin became American Olean.

Requirements of the automobile industry for higher-strength steels free of refractory inclusions have led to the use of higher quality refractories in contact with the melt. Traditional fireclay linings have been replaced by high alumina, silicon carbide and carbon. Silica or high alumina is now used in blast furnace stoves.

Steelmaking is now done mostly in basic oxygen furnaces (BOF) rather than open hearths. Magnesia combined with graphite or carbon is used as the refractory lining. The resistance of this lining in combination with new techniques of hot maintenance of the lining by flamegunning has extended furnace lives to as many as 10,000 heats. Maximum throughput of steel in the BOF has led to final adjustment of composition in the ladle, which again required replacement of the fireclay refractory with high alumina or magnesia. Improved slide gates made of high alumina, magnesia and alumina graphite have been developed.

In contrast to metals, crystalline ceramics are most commonly made by consolidating powders (sintering). Melt processing is also used for glasses, glass-ceramics, single crystals, and a small amount of so-called fused cast refractories.

The major stages in the powder-consolidation method of ceramic production are production of the powder, forming of the unfired article and sintering (firing). Traditional starting powders for ceramic processing are made from naturally occurring mineral deposits. Removal of iron by magnetic separation is important in making ceramics free from discoloration. Many high-technology ceramic products require synthetic powders. For example, alumina spark plugs depend on high-purity alumina prepared in bulk by processes developed mostly in the 1930s.

Early ceramic forming techniques were simple (although the resulting shape was sometimes complex) and depended on skilled hand labor. Traditional brick and tile were used as fired. In the 20th century complicated ceramic shapes with close tolerances have increasingly been required as well as methods of mass production. Machining of a fired ceramic is sometimes necessary but is held to a minimum because the high hardness of most ceramics makes mechanical finishing expensive. Accordingly, most ceramics are formed and fired to produce a shape as close to the final shape as possible, placing special requirements on the forming and sintering stages. Contemporary forming techniques include slip-casting, extrusion, uniaxial pressing, isostatic pressing, tape casting and injection molding. Injection

Personnel at a ceramic experiment station posed in 1921. R.T. Stull, standing fourth from the right, was president of the Society 1919-20. The station was part of a cooperative study on heavy clay products by the Bureau of Mines and several brickmaking associations. The creation of the Bureau of Standards in 1901 was a milestone in ceramic industry, as the Bureau did considerable basic research benefiting ceramic manufacturing and set standards for thermometers, thermocouples and optical pyrometers.

Plant and mine tours were important ways of gathering and exchanging ceramic knowledge before modern media linked industry worldwide. American ceramists visited European plants on a 1928 Society-sponsored trip.

At the Drakenfeld (now Cerdec Corporation) glass colors division and other companies in the glass industry, business boomed with the repeal of prohibition in 1933, which set off a tremendous flurry of activity to produce beer and liquor bottles, not to mention glass-lined brewing tanks. Glass makers also had a bonanza in the development by Corning of glass blocks, an essential element in the Art Deco design of the period.

For glass makers the biggest news was state legislation requiring the use of safety glass in automobiles. By the mid-1930s, 26 states required safety glass, setting off a rapid evolution as the industry turned from cellulose nitrate (which tended to brown and blister) to cellulose acetate (which remained relatively brittle) to the German-developed "Sekurit" and finally the forerunner of today's windshield glass, the Libbey-Owens-Ford development of a glass with a laminated vinyl plastic center.

molding was developed in the 1920s and in mass production in the 1930s. Isostatic pressing of ceramics for spark plugs was introduced in the 1930s and is the predominant manufacturing technique used worldwide for such items today.

Several major advances in the mass production of both traditional and high-technology ceramics took place in the 20th century with great progress being made in the 1930s through the 1960s. Improved methods of controlled agglomeration of powders, such as spray drying, facilitated rapid and well-controlled handling of powders. New forming techniques, such as dry-bag isostatic pressing, were introduced for mass production of spark plugs in the 1930s and were introduced into the manufacture of many other ceramic items. New, low-cost yet high-purity alumina, such as low-soda alumina, made possible mass use of alumina-based ceramics in many electronics applications where low electrical conductivity is required. New, sinterable powders such as alumina doped with titanium or other additives made sintering as low as 1400 C possible and made mass production of alumina-based ceramics less expensive and more practical. Advances in tunnel kiln technology included the development of the roller hearth kiln with ceramic rollers, improved burner technology for lower cost and more uniform heating, and high-efficiency, lightweight fibrous thermal insulation. Together these tunnel kiln advances revolutionized manufacture and use of tunnel kilns. Another innovation with large and still-developing applications is that of sheet forming of ceramics by the tape casting process. Cutting and stacking produces multilayer assemblies that can be metallized to make multilayer capacitors or interconnected in three dimensions to form compact and hermetically sealed wiring harnesses for electronic chips.

A major innovation of the late 1950s with large and continuing impact was the development of sheet-forming of ceramics (also known as tape casting). John Lawrence Park of American Lava Corp. extruded thin sheets of alumina and other ceramics with an organic binder and plasticizer and received a patent in 1961. The process was further developed by knife-coating the slurry onto a plastic film to create the tape casting process widely used today. The process has undergone many detailed improvements including the use of better binders and plasticizers and a surface roughness of one to two microinches that facilitates the application of conducting films. Patterns of electrically conducting lines were initially applied by screening-on pastes and later by sputtering films. Thousands of holes for location or pass-through conductors can be punched with a press.

Manufacturing automobiles required steel and other high-temperature products, which in turn encouraged the growth of ceramics designed for high-temperature work.

This technology was further advanced by Bernard Schwartz of IBM who conceived the idea of laminating these sheets, reduced it to practice, and described it in a 1961 paper. Two important classes of products resulted: multilayer capacitors and multilayer ceramic electronic packaging. Both are made in the tens of billions per year. This technology is still being extended to produce a new generation of products. Pioneering efforts include incorporation of portions of radio frequency wireless devices within the laminated layers.

Sintering (i.e., final densification of the formed shape) of traditional ceramics includes the formation of a liquid phase during firings at high temperature which subsequently quenches to a glassy phase at room temperature. A major achievement of the mid-20th century was the development of a science of sintering that recognized several possible forms of matter transport. These included liquid-phase transport and transport by diffusion through the solid. Sintering is today an empirical process whose optimization for particular ceramics has been assisted by the understanding provided by sintering science. Sintering in the absence of a liquid phase is required for many high-technology ceramics. Such ceramics typically have substantial remaining porosity.

Sintering science led to an understanding of how pores become trapped within grains and cannot be eliminated. A triumph of ceramic technology in the 1960s was the development of the Lucalox ™ process for sintering pore-free alumina and its extension to several other ceramics. Some covalent ceramics, such as silicon carbide, will not densify in the solid state under normal conditions. Another triumph was the development in 1973 of a process involving iron and boron additives for making dense silicon carbide.

The rising concern and increasing government involvement in safety, health and the environment in the 1960s had major impacts on industry including the ceramic industry. In the United States the Environmental Protection Agency and the Occupational Safety and Health Administration have been particularly important, as have their equivalents at the state level. Practices in the

Today a silica-coated aluminum nitride that is much more thermally conductive than standard molding compounds is used for a range of packaging applications. Its forerunners were in development during the 1930s when ceramists began turning from naturally occurring Silicates to the synthetic powders that later would be required by high-technology ceramics. In the next half century following, ceramists would work less and less with simple clays and come to rely on the purity and known quantities of synthesized ceramic materials: silicon carbide, silicon nitride, titanium boride and a host of others.

By the 1940s, ceramic materials were machine-formed to a myriad of useful shapes, such as these tiles made for schools and hospitals. But even so, World War II found ceramic industries at their most vulnerable. Ceramics were largely overlooked by the mobilization effort. Ceramic engineers were not deemed vital enough to the cause to warrant an occupational exemption, and as a result, the industry was decimated by the draft.

The war effort affected some consumer ceramics especially. The Supply Priorities and Allocation Board in 1941 banned all nonessential public or private construction which would divert vital materials from the defense effort. At the same time the Office of Production Management ordered a 43 percent cut in production of porcelain enamel, putting a dent in the home appliance market.

By 1942 the War Production Board had ordered an end to stove production as enamelers turned their attention to making war items such as shells and boats. At one point washing machine and refrigerator production was halted entirely. Meanwhile, glass manufacturers benefited from the war effort, with glass containers replacing cans for domestic packaging.

MURATA — INNOVATOR IN ELECTRONICS

Murata Manufacturing Co., Ltd. is an integrated electronic components manufacturer. Since its establishment in 1944, it has pursued research and development into a broad spectrum of functional ceramics and their potential applications, and has introduced a diverse range of electronic components that utilize the unique electrical properties of ceramic materials.

Functional ceramics are stone-like materials that are made by adding various chemical raw materials, separated and purified at atomic level, to a ceramic base that is hardened by firing. By adding trace quantities of doping materials and changing firing conditions or atmospheric conditions, these "wonder stones" can vary with a wide variety of electrical characteristics.

Over the years, Murata has created a rich variety of innovative electronic components with these "wonder stones." The ceramics technology is growing increasingly important as a key driving force in the creation of a new electronics revolution for the 21st century.

Murata believes that "new quality electronic equipment begins with new quality components and new quality components begin with new quality materials." Our work spans the range from basic scientific research through to the development of the next generation of advanced technology.

The Murata Group is committed to being an "Innovator in Electronics." Through our product innovations we hope to improve the quality of life in the world through our contributions to the development of this electronic society.

Founder, Akira Murata

handling of raw materials and manufacturing of ceramics have been modified in many cases to improve safety and health. Waste disposal has been modified to protect the environment. Ceramics have found many roles in improving safety, health, and environmental protection. Ceramic-wear materials protect many types of linings for high heat conditions, ceramic filters for removing contaminates, woven ceramic fibers for handling and filtering dust-laden waste streams and high-temperature bags for dust removal. Nonporous ceramic linings are used for collection and movement of gases and particulate matter. Porcelain-enameled metal sheets are used for corrosive gas exposure and enameled vessels for reaction and mixing vessels. Special cements line collecting ponds and hold together bricks in kilns and stacks. Some cements are designed for air and liquid exposure and offer excellent service life.

Early in the century (above), plate glass required an exacting polishing process, but by the late 1950s a new process eliminated grinding and polishing. In the decades that followed, America's cities sparkled as more and more glass was used in commercial building.

Advances in the glass industry have been central to many other industries. Major components of the glass industry deal with bottles, flat-glass, fiber glass and optical glass. In the late 19th century, many attempts were made to develop automatic bottle-making machines. Patents for various designs were issued in England beginning in 1859. In 1877 the semi-automatic Ashley machine was a commercial success. The first fully successful automatic glass bottle machine, the Owens bottle machine, was designed by M.J. Owens in the United States in 1898 and was in successful commercial production in 1904. A 10-head Owens machine could produce 2,500 bottles per hour. The individual section (I. S.) machine, which uses gobs of glass from a feeder, was introduced in 1925 and gradually superseded the Owens bottle machine.

The success of the electrical lighting industry in putting the Edison-Swan incandescent lamp into mass use depended upon the ability to make glass light bulbs cheaply. In the late 1920s the

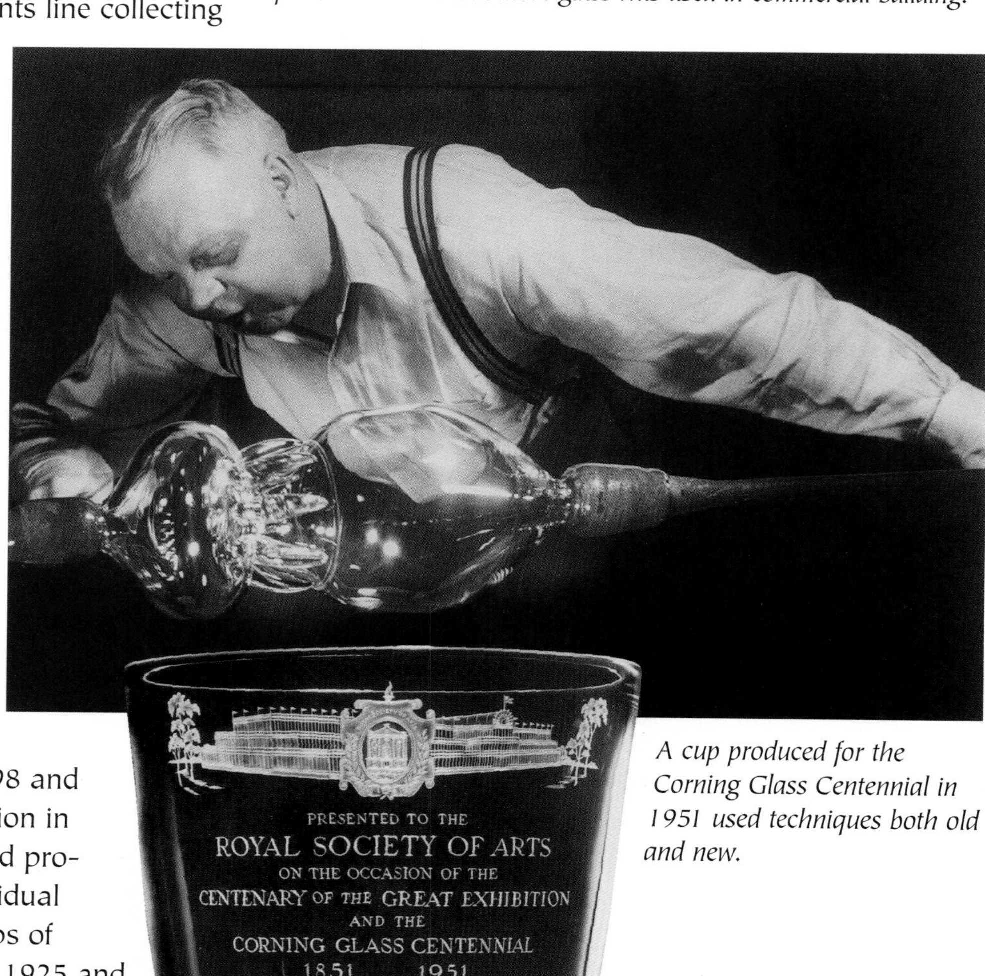

A cup produced for the Corning Glass Centennial in 1951 used techniques both old and new.

"No thanks — I'm not sure they aren't made of clay, too."

Technical advancement and corporate shifting marked the ceramic industry after the 1950s. An explosion of technology from the war effort made its way into the community at large as scientific data from government and industry research resulted in products and processes with myriad applications. Innovations reached into American homes with products such as glass-ceramic cooktops.

INDUSTRY SEES THE LIGHT

In the early years of the Society, the most divisive hurdle to be overcome was the reluctance — even outright refusal — of some commercial elements of the ceramic industry to allow their employees to participate fully and openly in the scientific and technical forums that were essential to the Society's goals.

Perhaps the most eloquent statement on this issue came from H.C. Wheeler, the first president of the Society, in his retirement address at the second annual convention in 1900. His words proved prophetic.

> *It has been predicted that our Society could not succeed, as many potters would not allow their technical men to attend the meetings, or if they did, that they would be under strict injunctions to keep their mouths shut, but their eyes and ears wide open.*
>
> *While I am aware that some of our most valuable workers are thus prevented by their employers from making contributions in the form of papers or discussions, lest supposed secrets or valuable trade information might be imported to their competitors, this is a sad blunder out of which our Society will have to try to educate them. For if the objections are keenly analyzed, they will be found to be relics of ignorance and unprogressiveness.*
>
> *For while some intellects are brighter than others, and the facilities of a few exceed that of many workers, no one man can hope to monopolize all the best things in any industry, and the ability of a congress greatly exceeds that of the individual . . .*
>
> *Though it requires time to educate the capitalist that by imparting information of a technical character he is investing brains and experience at compound interest, he will eventually find that his captains will receive from others far more than they give, and that all will be benefited by a broad, free interchange of experiences and criticisms.*

Corning ribbon machine was introduced. A continuous ribbon of glass is carried between opposed blowheads and blow molds. Glass bulbs are produced at a rate of 1,800 per minute. Fluorescent lighting requires glass tubes. In 1917 Edward Danner succeeded in making tubing continuously by flowing molten glass around a rotating, heated metal rod and drawing it from the end. Glass TV tubes of optical quality are made by a special centrifugal process.

Flat glass was made in the late 19th century by three processes. In one, molten glass was rolled flat in a frame, cooled and then ground and polished. In the second process, a cylinder was blown, cut along a line parallel to the axis, opened and flattened. In the crown method, a sphere was blown, opened and spun. In the 20th century a sequence of advances was made in forming sheet glass directly from the melt. The Fourcault process was patented in 1901 and operated commercially in 1913. Glass sheet is drawn upward from a melt through a slit and cooled. In the continuous plate-glass process of 1926, glass from a tank furnace flows to rollers. The resulting glass plate is then ground flat and polished. The next great advance in flat glass was the float glass process developed by Pilkington in the late 1950s in which molten glass flows from the tank onto a bath of molten tin and is then withdrawn on rollers. Grinding and polishing are eliminated.

Glass fibers are considerably more than a century old. Continuous-filament fiber glass is many centuries older than the glass wool type, but the latter was first made commercially in large quantities. Glass wool bats were used at least as early as 1840.

Many variants on the basic process of making glass wool using steam jets have been successful. Bats of glass wool came into large-scale production in the 1930s and are extensively used as thermal and acoustic insulation in the construction industry. Bats of glass fibers with compositions specially formulated for use at high temperatures can be used continuously at least as high as 2600 F. These insulating bats have revolutionized furnace design in the last 50 years. Air filters using glass fibers for the home furnace are ubiquitous. Fiber glass is also widely used in industry for liquid filtration.

Hand-drawn fibers were known in antiquity. At the Columbian Exposition in Chicago in 1893 a model even wore a dress made of glass fibers. However, mass commercial use of glass fiber melting developed greatly in the 20th century since the founding of the Owens-Corning Fiberglass Corporation in 1937 to mass produce

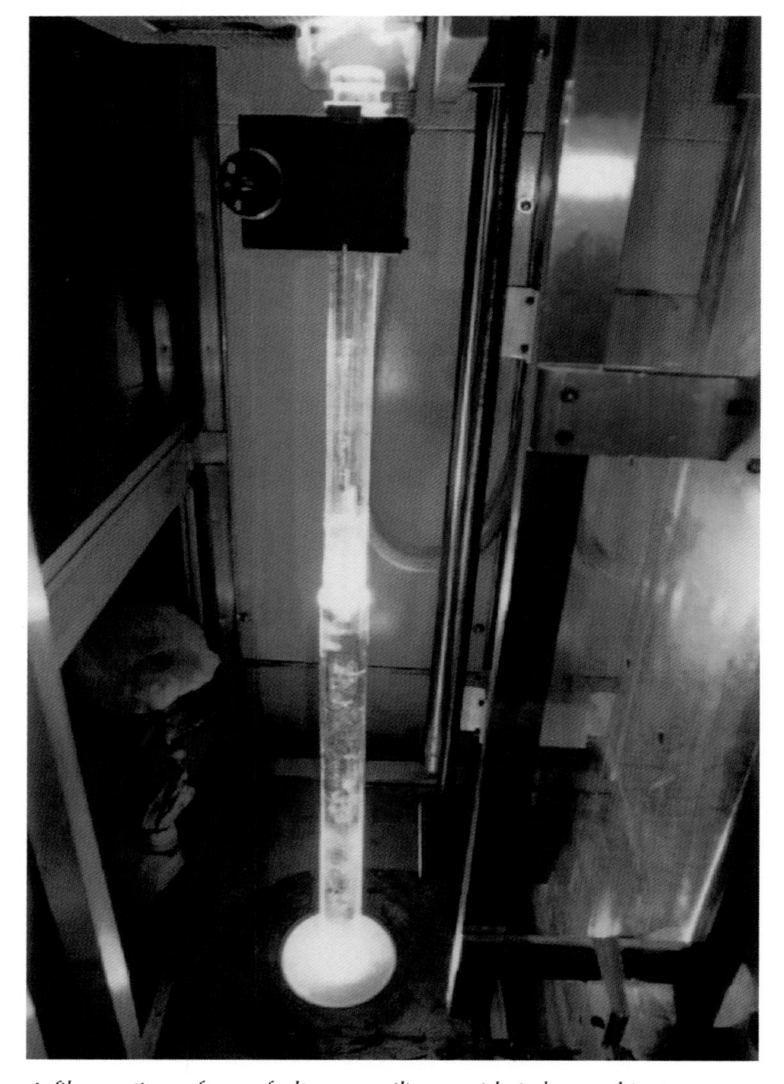

A fiber optic preform of ultrapure silicon oxide is lowered in to a 4000 F furnace for drawing into hair-thin optical fiber at Lucent Technologies' Atlanta Works manufacturing facility. Optical fibers are now being designed to carry up to 400 billion bits of information per second. This is done by simultaneously carrying 80 different colors of light. A single such fiber will carry 10 million simultaneous phone calls, the entire Internet's traffic or transmit 90,000 encyclopedia volumes in one second.

The advent of nuclear power — and the advent of ceramic nuclear fuels — led to development of such exotic ceramic materials as hafnium oxide, boron carbide and graphite. In 1959, the Bulletin *noted: "It is significant to the ceramic industry that the majority of the power reactors now under construction or planned will use ceramic fuel elements. Only four years ago ceramic fuels were receiving only nominal support, both in and out of government facilities." Ceramics would also play a vital role in creating nuclear waste glasses to encapsulate the radioactive isotopes that are a waste product of nuclear reaction.*

Today's semiconductor industry uses products like these contiguous slotted wafer boats for benefits such as higher temperature stability in a variety of applications. Initiated by the 1947 development of the transistor, semiconductor chip technology has spawned much research into the insulating, dielectric, piezoelectric and magnetic properties of ceramics, opening a whole new world for ceramic products.

continuous glass fibers. These fibers have been widely used as reinforcement in polymer-matrix composites and are found in many applications from boat hulls to aerospace vehicles.

Optical uses of glass fibers include both short-path and long-path communication. Bundles of continuous fibers are used over short distances for light and image transmission, including such uses in medical devices for diagnosis and surgery. High-silica optical fibers are now widely used for long-distance communication. A recent development is the fiber-optic amplifier, which amplifies light much as a laser does without the usual need of detectors, electronic amplifiers and laser transmitters.

Advances in ceramic technology in the last century have an enormous variety of applications. Following are some in which very important innovations have occurred, many of which cross and connect these descriptive areas.

ARCHAEOLOGY

Pottery was one of civilization's earliest innovations. Archaeologists have developed extensive tables of pottery styles and dates that are cross-correlated with other historic records to more accurately date the strata these scientists study. New techniques such as radiocarbon dating and thermoluminescence dating — the latter of which can be performed on tiny ceramic fragments — allow archaeologists to develop an ever clearer picture of man's past.

AUTOMOBILES AND RELATED VEHICLES

Although they are commonly overlooked, ceramic items can be found in many of the devices and systems in today's automobiles, trucks and specialty vehicles. From early innovations such as spark plugs and window glass to modern sensors that control engines, ceramic components are essential to the automobile industry.

In modern automobiles, ceramics — such as the honeycomb ceramic in catalytic converters — play an important role in pollution abatement. The ceramic oxygen sensor is today found on all auto engines in the United States and is an essential part of efficient and low-pollution engine operation. Ceramic components also improve a

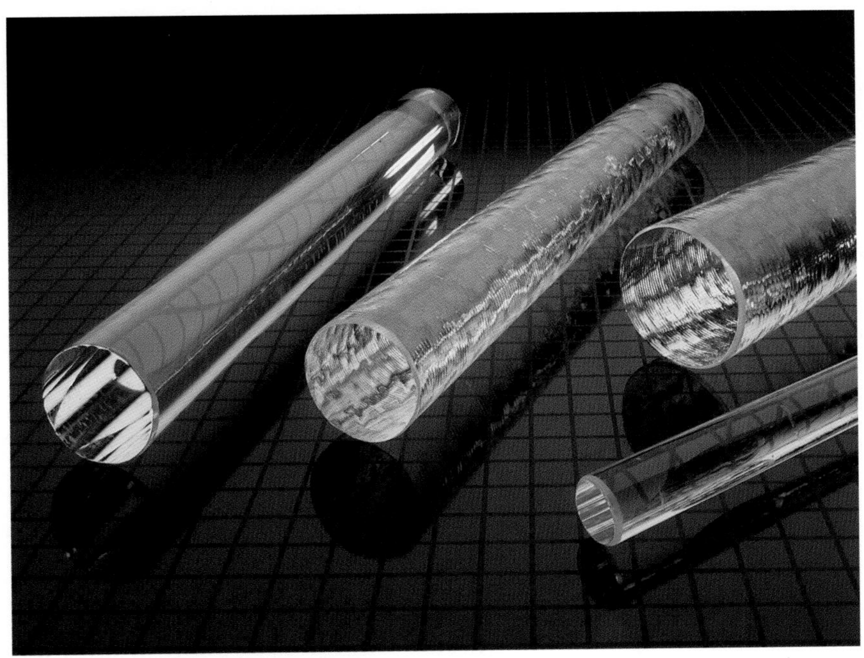

Glass for nontelecommunications has also found important applications. The rods shown here become the core of fibers that are used in such diverse applications as night-vision goggles, image guides, fiber bundles, endoscopes, and other instruments.

Continued on page 158

KYOCERA AND AVX CORPORATION: A PROFILE OF TWO INDUSTRY LEADERS

Kyocera and AVX Corporation salute The American Ceramic Society for its 100 years of growth and success! This year's celebration of "a century in ceramics" reflects a unique role that only ACerS can perform. From the late 19th century through today's Digital Revolution, no professional association has been more dedicated to the issues and opportunities that surround our industry. Why do we support The American Ceramic Society? A closer look at who we are and what we do says it all.

(Courtesy of Kyocera Corporation) Kyocera's structural ceramics.

KYOCERA CORPORATION

Kyocera (NYSE: KYO) is perhaps the world's leading producer of advanced ceramics, with approximately half of its $5.76 billion in fiscal 1997 revenue coming from ceramic-related products. The company has been led by a spirit of innovation and creativity in providing solutions for its customers. Our credo: "Do next what others tell us we can never do."

Despite this leadership position, however, Kyocera is not the industry's best-known player — and has in fact grown from modest beginnings as a start-up enterprise. Dr. Kazuo Inamori, founder and Chairman Emeritus, established Kyocera as the Kyoto Ceramic Co. in Japan in 1959 with seven colleagues and about $10,000 in venture capital. This was barely enough to rent part of a local warehouse as a "manufacturing plant" and begin seeking customers. Nonetheless, Inamori's faith in the future of ceramics led him to work ceaselessly on the most challenging technical problems — until a breakthrough would occur to him, he says, "like a whisper from God."

(Courtesy of Kyocera Corporation) Silicon nitride cam rollers for heavy-duty diesel truck engines (produced by Kyocera Industrial Ceramics Corp., Vancouver, WA).

Over the following 39 years, Kyocera has used its core competency in advanced ceramics to develop innovative, vertically integrated products in such fields as electronics, telecommunications, metal processing, automotive components, optics, medicine and energy.

In Mountain Home, North Carolina, the facilities of Kyocera Industrial Ceramics Corporation possess more than three decades of experience in producing ceramic components for the wire, textile and paper industries. This site is also one of the company's key production bases for ceramic and cermet cutting tools.

In 1992, Kyocera Industrial Ceramics Corp. established an Advanced Ceramics Technology Center in Vancouver, Washington as a production and development base for high-performance structural applications. This facility became the first in North America to produce and ship structural ceramic engine components in mass production volumes. Kyocera employees in Vancouver presently manufacture silicon nitride cam rollers for heavy-duty diesel truck engines and silicon nitride seal runners for jet aircraft engines, among other products.

(Courtesy of Kyocera Corporation) Metallized ceramic RF/Microwave packages for telecommunications equipment (produced by Kyocera America, Inc., San Diego, CA).

Kyocera is also North America's leading producer of ceramic semiconductor packages for computers and telecommunications. The company operates production centers for these products at Kyocera America, Inc. of San Diego, California, and at Kyocera Mexicana, S.A. de C.V. of Tijuana, Mexico. In addition, Kyocera America, Inc. markets a unique line of applied ceramic consumer products made by Kyocera Corporation of Japan — including ceramic cutlery and kitchen tools, ceramic scissors and tweezers, even ceramic-tipped ballpoint pens.

Kyocera appreciates the support and encouragement we have received over the years from The American ceramics industry and from ACerS — the oldest and most successful professional association of its kind.

Thank you for making the growth of our enterprise possible.

(Courtesy of AVX Corporation) AVX is a world leader in ceramic Integrated Passive Components (IPCs).

AVX

AVX CORPORATION

From our humble beginnings in 1922 as the Radiola Wireless Corporation, operating out of a tiny shop on 17th Street in New York City, through all the incarnations and acquisitions since then, AVX has been dedicated to providing our customers with products and service that are recognized as the best the industry has to offer.! Today we design and manufacture passive electronic components with sales in excess of $1 billion annually. Our service ethic is highly regarded the world over. AVX is a partner our customers can rely on for quality products, service and delivery, and we are very proud of that.

Our tie to the ceramic industry is a significant one. AVX is the largest supplier of ceramic leaded capacitors in the world, and we have the largest ceramic capacitor manufacturing facility in North America ... right in beautiful Myrtle Beach, South Carolina. In fact, Myrtle Beach, our corporate headquarters, is the home of many of the world's breakthroughs in ceramic passive component technology. You'll find the AVX logo on ceramic products such as MLC capacitors, transient voltage suppressors, inductors, high-voltage capacitors, feedthrough filters, and numerous other devices.

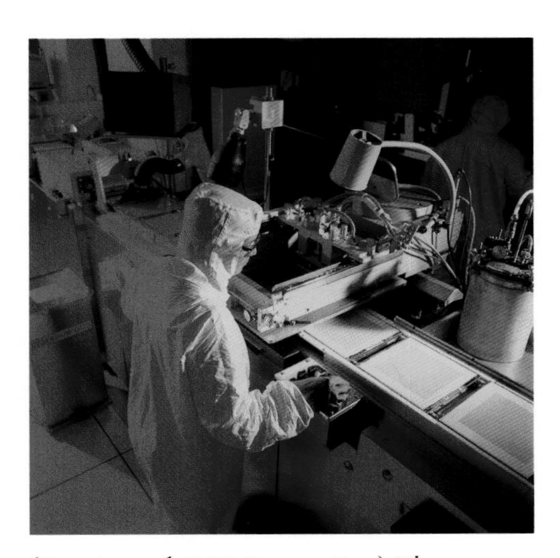

(Courtesy of AVX Corporation) The very latest ceramic capacitor manufacturing techniques with layer thicknesses of 3 microns and below, require clean room environments and processes.

Along with our ceramic component business, AVX is a world-class producer of tantalum capacitors and other passive components used in virtually every electronic product and application that stores, filters and regulates electrical energy. AVX also manufactures and distributes electronic connectors and distributes certain passive components made by Kyocera Corporation of Japan. AVX components play an integral part in today's sophisticated and fast-paced applications, including telecommunications, computers, automotive, medical and consumer electronics.

AVX employs more than 13,000 people in manufacturing, facilities and sales offices across Europe, North America and the Far East, and we produce parts in nine different countries. Our manufacturing capacity has grown to a global production level of more than 200 million components per day. As you might expect from an electronics innovator, AVX continues to invest heavily in research and development to position ourselves to meet the next generation of passive component requirements. Like The American Ceramic Society, we want to be the leader in our industry.

Continued from page 155

variety of engine parts. Diesel engines sometimes use silicon nitride precombustion chambers and exhaust valves. Ceramic seals are often used in water pumps. Metal-matrix composites employing ceramic fibers are coming into use for pistons and crankshafts. Ceramics even add convenience. Small electric motors with hard ferrite magnets — as many as 20 are used in some models — are found in applications such as power seats and door locks.

The vast and complex system of roads, signs and traffic control devices that regulate today's traffic employ ceramics in a variety of ways. In addition to concrete for road surfaces and bridge structures, road signs visible in all weather conditions are made possible by a polymer-glass composite using glass microsphere reflectors.

By the 1960s, consumer goods were flying off the shelves. Mass production of TV tubes was made possible by Corning's development of centrifugal casting methods. And glass cookware revolutionized the kitchen. Here Corning's Vision cookware holds a melting aluminum pot.

CERAMIC PROCESSING

Clay-based ceramics that employ naturally occurring fine particles date to antiquity. In the last century, new techniques that bypass the use of powder consolidation and improve traditional powder-based processing have been a major area of ceramic innovation. Magnetic separation is one such innovation. This process removes iron-containing particles that can cause discoloration from raw material, improving the composition of mineral deposits that traditionally were used with little or no benefaction.

Powder-based ceramic processing has advanced with the development of free-flowing powders that facilitate automatic mold loading and the development of controlled agglomeration to form free-flowing powder through the spray-drying process. Innovations including gel casting, enzyme catalysis in ceramic forming, pressure-assisted slip casting, isostatic pressing, and injection molding have improved the forming of unfired ceramics. Sol-gel processing, a technique that bypasses the powder stage, is now used for sol-gel fibers, abrasives and glass spheres. And new techniques for using applied pressure to facilitate sintering — including hot pressing and hot isostatic pressing enzyme catalysis in ceramic forming — are available. Synthetic powders (for example, silicon nitride powder improvement) are also now in wide use.

Once the province of sole proprietors and closely held corporations, ceramic companies in recent years have faced tremendous competition from other materials cutting into what had once been strictly ceramic markets. By the 1960s, the industry was shifting from one of individual owners to one of professional management teams working with skilled research and marketing staffs.

A boom in fiberglass, initially used mostly for insulation and filtration, reflected America's infatuation with boating. Today it is often used as reinforcing in polymer-matrix composites; it can be found in the hulls of both recreational boats and the space shuttle.

Automatic machinery for ceramic firing advanced notably in the last half century. The roller hearth kiln made possible rapid firing — as fast as 20 minutes — of ceramic tile and many other ceramic items. Conveyor technology for tile firing in tunnel kilns and advances in kiln burners were other important 20th-century ceramics innovations.

Glass-free ceramics traditionally were impossible to produce without some porosity. That changed with the development of pore-free ceramics such as alumina for sodium vapor lamp envelopes and the later development of pore-free silicon carbide ceramics.

CERAMIC COATINGS TECHNOLOGY

Ceramic coatings are used for a variety of purposes, including wear resistance, erosion resistance, thermal protection, control of optical properties and electrical insulation. Important innovations in this area include flame-sprayed ceramic coatings, thermal control coatings, sol-gel ceramic coatings, oxide infrared reflective coatings for lamps and windows.

CRYSTAL STRUCTURE OF CERAMICS

Ceramic properties depend not only on their chemical composition but also on their crystal structure and their microstructure. Two 20th-century developments that have improved ceramic structural analysis are X-rays in materials development and phase diagrams for ceramists. Although not products themselves, these innovations are important developments in the field of ceramics.

CONSUMER PRODUCTS

Ceramics are used in a wide variety of consumer products. Glass-ceramic cookware is familiar to everyone. Electric stove heating elements require a ceramic insulating layer to operate. Glass tableware, oxide non-reflective coatings for lamps, low-lead release glazes and lead-free glazes are some of the many ceramic products found in the average home.

The development of jet engines placed new demands on materials and greatly stimulated innovations in ceramics. Designers recognized the desirability of ceramic parts useful for their low wear, high hardness and high stiffness. Synthetic abrasives were developed for grinding and polishing super-tough metals and ceramics. Eventually, ceramic composite materials will begin to replace metallic super alloys in jet engines because of their lighter weight and higher temperature performance.

DEFENSE

Defense needs and opportunities have driven many advances in ceramics. In the quest to develop impenetrable defenses, ceramic armor plays an important role as an adjunct to metal armor. Piezoelectric ceramics have many applications, but their development has been pushed especially by the need for improved ultrasonic submarine detection using piezoelectric transducers operating in the kilohertz range. Following the development of barium titanate in the 1940s by Wainer and Salomon, lead zirconate titanate (PLZT) was discovered at the National Bureau of Standards by B. Jaffe, R.S. Roth and S. Marzullo in 1954 and remains the basis of most high-performance piezoelectric transducers. Smart, heat-seeking missiles require infrared detectors such as uncooled infrared cameras.

Educating consumers became important to ceramics, as this exhibit in an Ohio State Fair in the mid-1960s suggests. Ceramics organizations of all kinds looked to industry-wide programs for marketing, research and education to increase members' opportunities. The Society-sponsored expositions at annual meetings became an important center of shared energy and information to tackle the challenges modern ceramics faced in the marketplace.

ELECTRICAL AND ELECTRONIC USES OF CERAMICS

The insulating, dielectric, piezoelectric, magnetic and other properties of various ceramics have led to their widespread use in electrical and electronic devices. Electrical insulators (glass or porcelain) were used on early telegraph lines. Today multilayer integrated ceramic packaging technology rests on tape casting of ceramics and on metallization of ceramics that have been developed so that virtually any ceramic can be laid down as a thin film and that thin-film conductors can be applied and cofired with the ceramic. An intricate multilevel, three-dimensional composite of insulating ceramic and conducting metal can thus be produced to serve as a mechanical mount, a wiring harness and a hermetic seal for electronic chips in many applications.

Functional elements in electronic devices include capacitors with barium titanate dielectrics. High-frequency radio communication requires microwave dielectric ceramics that are used as components of highly selective filters for cellular telephones and satellite communication systems. Soft ferrites are required for small but sensitive radio antennas, transformers and inductors. Hard ferrites are used in many small electric motors including extensive use in automobiles as noted earlier. Zinc oxide varistors have become essential protection for electrical and electronic

A new growth area for ceramics is in biomedical applications. While ceramic dental prosthetics have been common for decades, in recent years ceramic bone and joint replacements have proven their value for toughness and resistance to the corrosive substances in the human body. Even more exciting is the possibility of new bone grown from a ceramic base. The new material Collagraft serves as a scaffolding on which a patient's bone cells can grow. Over time the ceramic materials are slowly dissolved by the body and replaced by natural bone.

equipment against voltage surges ranging from small overvoltages in electronic chips to huge surges from lightning strikes on power lines.

The discovery of ceramic superconductors was a very exciting development of the 1980s. These materials are now used in composites for magnet coils and in very low-loss microwave resonators for cellular telephones.

Although the ceramic boom of the 1980s has slowed, to a great degree because the end of the Cold War also signaled a precipitous decline in military-funded research, consumer products such as automobiles continue to provide incentive for research. In a Toyota research facility, a spin test (rotating at 85,000 rpm with a tip speed at the burst of 600 m/sec) bursts a radial rotor for a gas turbine engine made of Si_3N_4.

FILTERS

Ceramics play a central role in important types of filters. Fiberglass filters for dust filtration are found in every home ventilating and air conditioning system. Hot metal filters consisting of ceramics with controlled porosity are increasingly used to remove particles from metal melts.

GLASS PRODUCTION TECHNOLOGY

Many glass items are made in enormous numbers at very modest cost by automatic mass production methods. Innovations in glass manufacturing technology include the Corning ribbon machine for making light bulbs and the float glass process for making flat glass. Glass fiber making techniques include the die process for continuous glass fibers and the steam blowing process for glass fiber mats.

In the early 1950s Project Tinkertoy at Nation Bureau of Standards developed automated techniques for fabricating miniature circuits on stacks of ceramic substrates. By 1958 monolithic integrated circuits on diffused single crystal silicon were possible. The computer age was at hand and with it entirely new applications for ceramic technology as critical hardware components. Today computers have become important to research, design, and manufacturing operations in virtually every arena of ceramics.

MACHINERY

Ceramics are useful for low-wear, high-hardness and high-stiffness applications. A growing application is in hybrid bearings. Glass-to-metal seals are in widespread use to provide insulated access for electrical lines through metal walls. Extremely high-strength parts can be made of transformation-toughened ceramics and fiber-toughened ceramics.

MACHINING OF METALS AND CERAMICS

Machining of metals and ceramics requires cutting and grinding tools of greater hardness and wear resistance than the material being worked. Metal turning is done with hardened tool steel or with cobalt-bonded tungsten carbide for the most part, but for specialized applications ceramic cutting tools are used. Finishing of the harder metals is often done by abrasive grinding and polishing. The development of synthetic ceramic abrasives includes the synthetic superhard materials.

MECHANICAL PROPERTY ENHANCEMENT

Ceramics can be made quite strong but are characteristically brittle. Mechanisms for imparting greater toughness to ceramics have been extensively studied and some in particular have been quite successful. These are transformation-toughened ceramics, whisker-reinforced ceramics and fiber-reinforced ceramics.

MEDICAL AND DENTAL APPLICATIONS

Ceramics are increasingly used in medicine in diagnosis and as implants. Bioceramics are used as bone replacements, especially in Europe. Ceramic coatings are used on metal implants to facilitate bone growth; hydroxyapatite coatings are used for dental and orthopedic implants. Ceramics are very widely used in dentistry. Few older people are without a crown or bridge, and some 80 percent of these dental restorations are made of porcelain fused to metal.

In 1935 at a Kentucky-Tennessee Clay facility, mining operations were carried out with the help of a small steam engine and a narrow-gauge railway.

The introduction of the U.S. space shuttle program turned attention to ceramics used as linings to protect vehicles from the tremendous heat generated by re-entry into the Earth's atmosphere.

Ceramics was a glamour field during the 1980s, with such high-tech applications as cable-optic fiber and the ceramic heat shields for the U.S. space shuttles getting reams of publicity. Industry got a boost from Reaganomics, and the federal government lavished money on the ceramic research that many in government saw as the solution to numerous military and economic challenges. Particularly exciting was the discovery of superconductors, still one of the most intense areas of ceramic research.

Marvelously versatile and open to innovation, ceramics always present the possibility of being engineered in the shape of things to come, whatever forms that may take.

Computer-assisted X-ray tomography (CAT) has become an essential medical diagnostic tool. A recent advance in this technology is ceramic scintillators for medical X-ray detectors in CT-body scanners. Another widely used medical diagnostic tool is ultrasonic imaging. Ceramics are used for the signal generators and pickup; improved performance is provided by megahertz composite dielectric ceramics. Numerous types of medical instruments for body cavity inspections utilize fiber optic bundles.

NUCLEAR APPLICATIONS

Two major types of nuclear fuel involve ceramics. The majority of nuclear reactors throughout the world today use uranium dioxide as a nuclear fuel. Carbide nuclear fuel has been developed but its widespread use awaits the deployment of high-temperature, gas-cooled reactors. The planetary probes to Mars and beyond would not be possible without lightweight, long-lived and efficient power sources. This power is provided by ceramic fuel for space exploration utilizing plutonium dioxide. Nuclear power produces highly radioactive waste that must be stored for several half lives of the most radioactive isotopes. Nuclear waste glasses provide the means to immobilize these isotopes safely.

OPTICAL COMMUNICATION

Optical communication over long distances has been made possible by the development of fiber optics for communication with very low optical attenuation per kilometer. Even with such low-loss fibers the signal must be amplified periodically. This amplification was done at first by detecting the optical signal, amplifying it as an electronic signal and retransmitting it with a modulated light source. The recent development of fiber-optic amplifiers has made it possible to amplify the light itself.

Important defense applications for ceramics continue in the late 20th century. For example, new, reinforced ceramic armor (left) can stop heavy machine gun rounds and even take additional strikes that would have shattered conventional armor.

RADIO AND TELEVISION

Soft ferrites provide cores for high-sensitivity reception antennas and shape magnetic fields to deflect the electron beam in the picture tube. Television tubes require excellent optical quality in the front face and

Continued on page 166

CORNING INCORPORATED — CREATING THE WORLD'S LEADING-EDGE TECHNOLOGIES

From the glass for Thomas Edison's electric light bulb to optical fiber — the heart of today's communications revolution — Corning's research and development efforts have yielded life-changing and life-enhancing inventions, which have made possible entirely new industries.

Today, Corning's worldwide R&D efforts continue to shape our world and are the driving force behind businesses that will lead the company into the 21st century.

(Above,) Optical fiber, invented by Corning in 1970, is at the heart of a global revolution in communications. Corning also is a world leader in photonic components that vastly enhance the effectiveness of fiver communication networks.

(Left) Glass for liquid crystal display screens made with a Corning process that creates the flattest, purest glass in the world.

(Above) Cellular-ceramic substrates for catalytic converters supplied to all of the world's major automakers.

An unwavering commitment to technological innovation has built a rich legacy of scientific achievement at Corning. Innovation is part of Corning's genetic code.

(Below) Lenses made from high-purity fused silica used in the manufacture of the next generation of semiconductors.

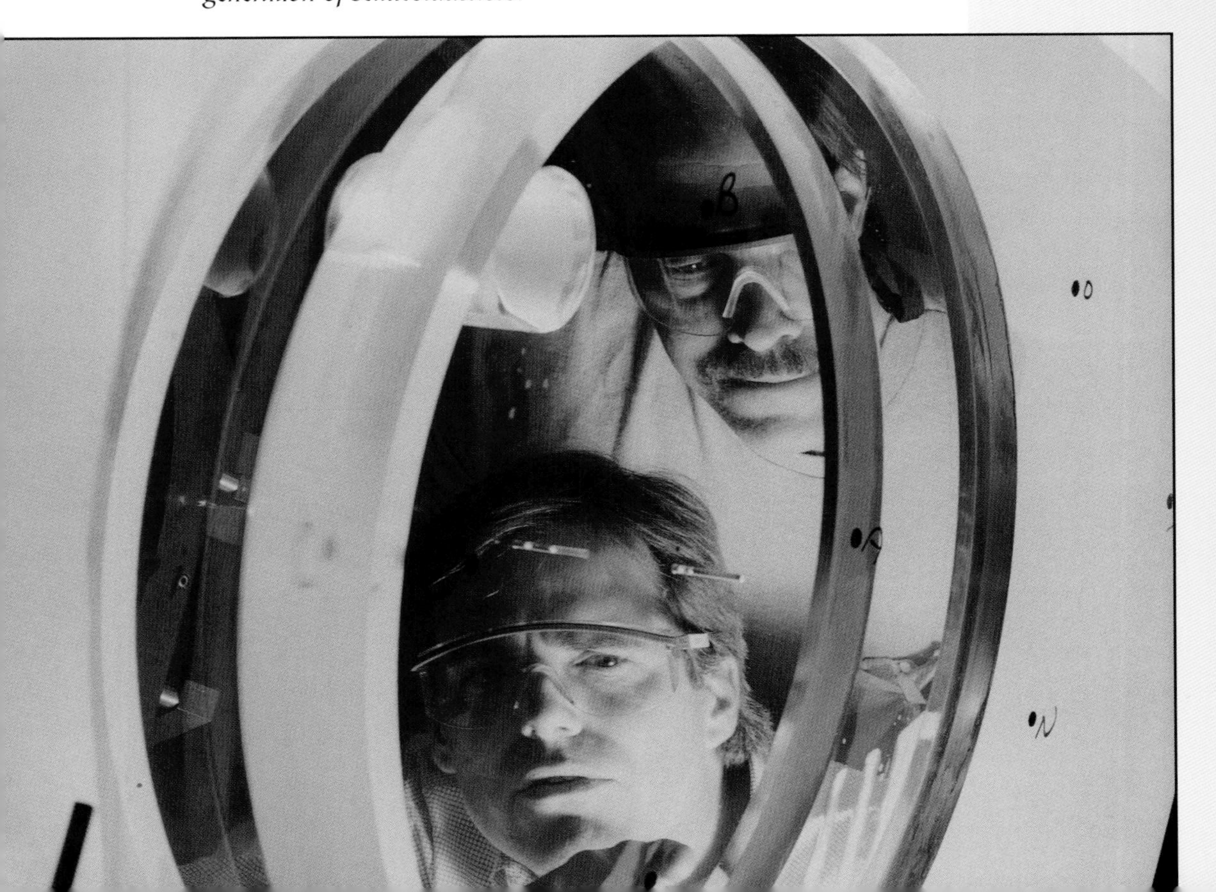

Continued from page 163

high dimensional control. The development of mass-production methods for making optical quality TV tubes has been essential to the vast worldwide use of TV.

REFRACTORIES

Refractories for steelmaking have advanced continuously but at least three major innovations can be recognized. These are the castable refractories, magnesia-carbon refractories for the basic oxygen furnace and the refractory slide gates for ladle metallurgy. Refractories for general use are often subject to thermal shock. The development of new low-expansion ceramics minimizes this problem.

SENSORS

A wide and growing range of ceramic sensors is used, often in applications for defense, automobiles, electronics and optical communication. Others include positive temperature coefficient resistors and fiber optic sensors.

SINGLE CRYSTALS

The ability to grow large ceramic single crystals such as synthetic sapphire and ruby has led to many applications including watch bearings and jewelry. Large ceramic single crystals are also used as lasers, including ruby, Ni-doped YAG and Ti-doped sapphire. The latter is used with an argon pumping laser to provide a tunable laser.

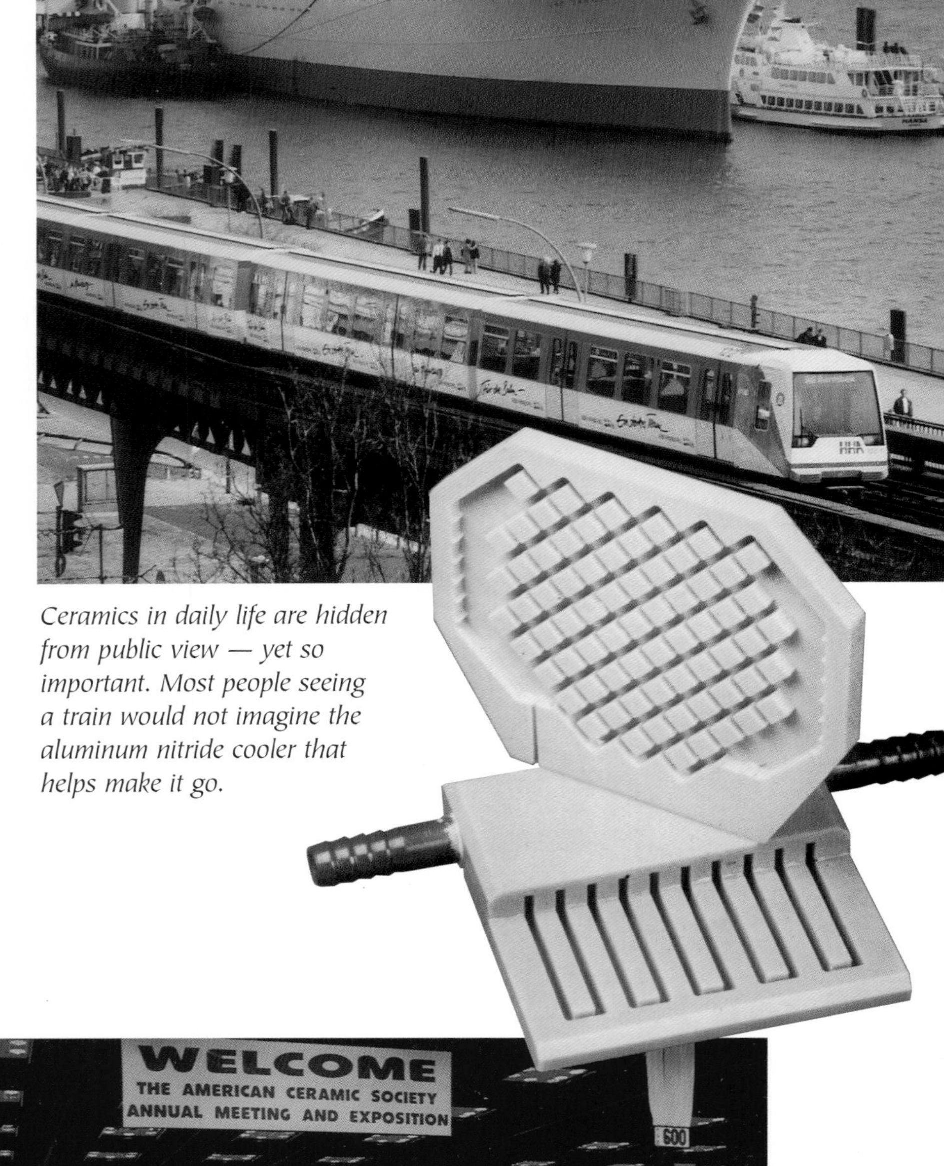

Ceramics in daily life are hidden from public view — yet so important. Most people seeing a train would not imagine the aluminum nitride cooler that helps make it go.

SYNTHESIS OF TOTALLY NEW CERAMIC MATERIALS

The important ceramic material silicon carbide was unknown until it was discovered during attempts to make diamonds. The Acheson process for silicon carbide made possible a large industry producing cutting, grinding and polishing wheels and saws. A variety of synthetic materials are made for special functional requirements, including barium titanate and lead zirconate titanate.

In 1982, Roy, Alamo and Agrawal reported the discovery of the largest structural family of materials with tailorable thermal expansion from modestly negative through zero to low positive values. This family was

At the 1989 Ceramics Exposition, ACerS President William Rhodes (right) admired the Nielsen racing car, a Chevrolet Camaro stock car with ceramic intake and exhaust valves.

Today even sports equipment has a "mind" of its own. A peizoelectric vibration control module is a ceramic application that made "smart skis" a winner in Popular Science *magazine's "Best of What's New."*

named "NZP," representing the parent compound ($NaZr_2P_3O_{12}$). The [NZP] structural family was found to consist of more than 150 members, many of which have been characterized by their ultra-low thermal expansion behavior. This perhaps is a unique structural family in which, by careful ionic substitution, one can develop a composition of virtually any thermal expansion value over a wide temperature range suitable for a specific application. The [NZP] family permits a variety of engineering possibilities, such as devices requiring coating materials for carbon composites, and recently, active catalysts for NOx reduction.

Even a brief look at ceramic innovations over the last century reminds us that the applications for ceramics known today are just the next phase in a vast, expanding continuum of scientific discovery and technological application. ▲

1894 The first ceramic engineering course is established by Edward Orton, Jr. at The Ohio State University.

1898 During the National Brick Manufacturers' Association meeting in Pittsburgh, (February 15-18) the formation of a new ceramic association is discussed.

1899 The "First Meeting" to organize a permanent organization is held in Orton Hall with Edward Orton, Jr. as secretary.

1899 The First Summer Excursion Meeting takes place July 4-7, with a session aboard the *Put-In-Bay Steamer* en route to Put-In-Bay, Ohio.

1899 The first volume of *Transactions of the American Ceramic Society* is published.

Edward Orton, Jr.

1900 The New York State School of Clay-Working and Ceramics is established and run by founder Charles Fergus Binns.

1901 The first associate members are elevated to the grade of full members at the Third Annual Meeting.

1902 Beta Pi Kappa, the ceramic engineering fraternity, is established at The Ohio State University.

1902 *The Collected Works of Hermann August Seger* is translated and published by the Society as its first major project.

1902 Rutgers University establishes its Ceramic School.

1902 The Summer Excursion Meeting takes place September 5-8 at the World's Fair in St. Louis, a "wholly social" event.

1905 The Society's Articles of Incorporation are subscribed, acknowledged and filed in the office of the Secretary of State in Ohio.

1908 The first Society banquet is held February 5 at the Hartman Hotel, during the Annual Meeting in Columbus, Ohio.

(Opposite) This ceramic art piece, Masked Man Cup, *by Wesley Anderegg, was the winner of the Ceramic Arts Contest in 1998. The competition was held in honor of The American Ceramic Society's 100th year.*

100 EVENTS TO REMEMBER

1914 The Ohio State University's application to form the first student chapter of the Society is approved at a Society meeting February 26.

1915 The first local section of the Society is founded by a group in Beaver Falls, Pennsylvania.

1915 Keramos is founded.

1916 The Society decides it is no longer necessary to hold its convention in conjunction with National Brick Manufacturers' Association meetings.

1917 Charles F. Binns succeeds Edward Orton, Jr. as Secretary of the Society.

1918 The *Journal of the American Ceramic Society* publishes its first volume, replacing the annual *Transactions*.

1919 Professional divisions are formed at the 21st Annual Meeting, held at the Fort Pitt Hotel in Pittsburgh, February 3-6.

1919 For the first time, paper abstracts are published as part of the *Journal*.

1920 The first separate sessions for divisions (Enamel, Glass, Refractories and Terra Cotta) are scheduled at the 22nd Annual Meeting February 23-26.

Ross C. Purdy

1921 Ross C. Purdy is named the first full-time General Secretary of the Society, a position he would hold until 1946.

1922 All divisions hold separate meetings at the 24th Annual Meeting, February 27-March 3, because the Society has "so increased its scope."

1922 The *Ceramic Abstracts* are published independent of the *Journal* for the first time.

1922 Volume 1, Number 1 of the *Ceramic Bulletin* is published in May.

1922 The Society's editorial offices in Illinois and business offices in Columbus, Ohio, combine and relocate in new quarters on The Ohio State University campus in Columbus.

1925 The Ohio State University celebrates the 30th Anniversary of the founding of Ceramic Education.

1925 The Ohio State University ceramic engineering students are installed as the Beta Chapter of Keramos.

1925 Ceramic Day is observed at the Chemical Exposition Fall Meeting.

1925 The first high school ceramic courses are taught in East Liverpool, Ohio.

1926 The Society's offices move to 2525 North High Street in Columbus, Ohio.

1928 The Society holds its Summer Tour to Europe May 19-July 5.

1929 The week of the 31st Annual Meeting, held in Chicago, February 4-9 is designated National Ceramic Week and the "First Exposition of Ceramic Products in America."

1930 The Glass Division becomes the first division to hold a fall meeting, at Cove Point, Maryland, October 4-6.

1930 The first One Hundred Fellows holds its Organizational Meeting.

1930 Edward Orton, Jr. becomes President of the Society.

1931 At the first induction ceremony in Cleveland, on February 25, 153 members are inducted as Fellows.

1933 The Summer Excursion meeting is held in Chicago, during the Century of Progress "Engineers' Week," June 25-30.

1933 The first Edward Orton, Jr. Memorial Lecture is given by E.W. Washburn on February 14.

1933 *Phase Diagrams for Ceramists*, compiled by F.P. Hall and Herbert Insley, are published for the first time in the October issue of the *Journal*.

1935 The Association of Ceramic Educators is organized.

1936 Sessions are split between two hotels for the first time, at the 38th Annual Meeting in Columbus, Ohio, March 29-April 4.

1937 Engineering curricula are accredited for the first time.

1938 Formation of classes is approved by the Board of Trustees.

1943 The first "double the membership" or "member get a member" campaign is begun.

1946 The Purdy Collection of ceramic objects is given to the Society.

1946 Ross Purdy retires from his position as General Secretary of the Society.

1946 Charles S. Pearce is named General Secretary and serves until 1963.

1953 The publication of *Ceramics Monthly* begins in January.

1954 The first Society-owned headquarters building, in Columbus, Ohio, is dedicated December 4.

1957 The first Ceramographic Exhibit is shown in Dallas.

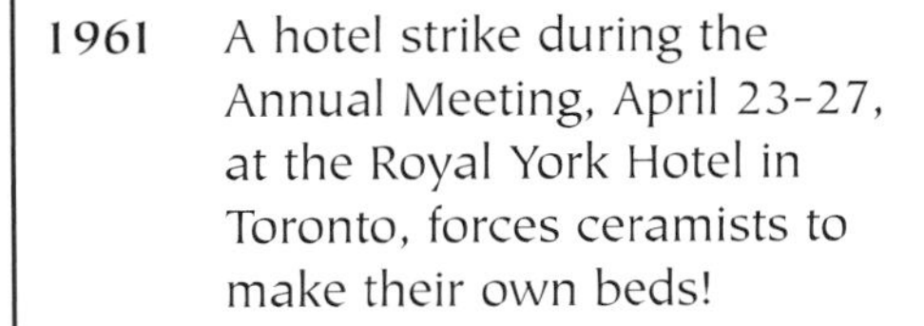

1961 A hotel strike during the Annual Meeting, April 23-27, at the Royal York Hotel in Toronto, forces ceramists to make their own beds!

1963 Frank P. Reid is named as General Secretary and serves until 1979.

1964 The *Ceramic Company Directory* is published for the first time.

1966 The Design Section of the Ceramic Education Council votes to split from the Society and form the National Council on Education for the Ceramic Arts (NCECA).

1967 An addition to the Society's Headquarters Building is dedicated June 1.

1968 For the first time, paid Society memberships and publication subscriptions exceed 10,000.

1969 The First Annual Exposition is held at Washington, D.C.'s Sheraton-Park Hotel, May 4-6.

1969 The Constitution is thoroughly revised and updated.

Charles S. Pearce

Charles S. Pearce joined the Society staff in 1944 as associate editor of publications and was named general secretary and chief administrative officer in 1946, a post he held until retirement in 1963. In those years, the Society enjoyed unprecedented growth and expansion, from 3,070 members in 1946 to 6,200 members at the time Pearce retired. The number of local Sections increased from 10 to 25, and two new Divisions (Basic Science and Electronics) were created.

During his tenure as general secretary, he also served as the editor of all ACerS publications. In recognition of his contributions, he received the Presidential Citation of The American Ceramic Society at the 65th Annual Meeting.

Born in Wichita, Kansas, Pearce later attended the Carnegie Institute of Technology and received his bachelor's degree in business administration from The Ohio State University. His years of working in the ceramics industry were mostly spent with one company, Delco-Lite Co. in Dayton, Ohio, which would later become the Frigidaire Division of General Motors. Beginning in 1926, he held a range of supervisory positions with Frigidaire including managing the porcelain enameling department, then the largest wet-process enamel plant in the world.

Pearce was named a Fellow of The American Ceramic Society in 1947. His other professional affiliations included a position as trustee and treasurer of the Edward Orton, Jr. Ceramic Foundation and both the presidency and vice-presidency of the Council of Engineering Society Secretaries. He died in 1972.

Frank Reid

Frank Reid was named general secretary of the Society in 1963 and held the position until 1979. At that time, the title of the position was changed to executive director. Prior to his term as general secretary, Reid had held several positions with the Society: business manager (1952-1954), assistant secretary (1954-1959), then associate secretary (1959-1963.) He also served as publisher of the Society's three periodical publications at that time.

During Mr. Reid's tenure as executive director, he oversaw the expansion of member services, including the addition of an exposition to the annual meeting, and membership promotion.

Before his years with The American Ceramic Society, Mr. Reid attended The Ohio State University, then served with the United States Air Corps. He later completed special courses in association management.

He was a trustee for the Edward Orton, Jr. Ceramic Foundation and was also a Fellow of the Society. In addition to his work with the Society, Reid was a member of Keramos, The American Society of Association Executives, and numerous other professional organizations. He was also actively involved in organizing Pilot Dogs, Inc., a national non-profit organization that trained guide dogs for the visually impaired.

He received the Albert Victor Bleininger Memorial Award in 1975. In 1977, on his retirement, Frank Reid also received the title of Honorary Life Member, the highest honor awarded by The American Ceramic Society.

Mr. Reid died in 1996.

1970 The Ceramic Endowment Fund is established.

1973 The new Society-owned Ceramic Park office building is completed.

1975 The first CEC short course is offered: "Kinetics in Ceramic Processes."

1978 The first Poster Session is held at the 80th Annual Meeting in Detroit, May 6-11.

1979 Arthur Friedberg succeeds Frank Reid as Executive Director and serves until 1984.

Arthur L. Friedberg

Arthur L. Friedberg was executive director of The American Ceramic Society from 1979 until his untimely death in 1984. A member of the Society since 1946, he was made a Fellow in 1956. He also served as a trustee of the Society's Ceramic-Metal Systems Division.

Remembered for his caring, involved management of the Society, Arthur Friedberg brought the perspective of an educator to the job. He worked with particular intensity on the Society's publishing and programming efforts.

He was an educator with a long and distinguished career. After earning a bachelor's of science in ceramic engineering from the University of Illinois, he attended the University of Chicago as well as the Scripps Institute of Oceanography at La Jolla, Calif. He worked for Stone and Webster Engineering Corp., and the Elwood Ordnance Plant in Joliet, Ill. From 1943 to 1946, he served as an officer in the U.S. Navy.

Friedberg received a Ph.D. in 1952 while serving on the University of Illinois teaching staff and became the head of the Department of Ceramic Engineering at the University of Illinois in 1963. A registered professional engineer, he worked on numerous research projects, wrote technical papers and consulted on ceramic technology. He was also a member of the Edward Orton, Jr. Ceramic Foundation's board of trustees.

1980 The Board of Trustees affirms action by the Publications Committee to establish *Communications*.

1980 The first issues of *Ceramic Engineering and Science Proceedings* are published.

1980 The first edition of the *Advances in Ceramics* series is published.

1980 A Society delegation goes to the People's Republic of China.

1981 The Ross C. Purdy Museum of Ceramics in Columbus, Ohio, is dedicated and opens.

1981 ACerS is adopted as the official acronym of The American Ceramic Society.

1981 A Society delegation goes to the 90th Annual Meeting of the Ceramic Society of Japan.

1982 On December 21, the Society and the National Bureau of Standards sign documents launching the Joint Program on Phase Equilibria for Ceramists.

1982 A Society delegation goes to Brazil for the Brazilian Ceramic Association annual meeting.

1984 The Society sends a delegation to visit the Ceramic Society of Japan, May 12-26.

1985 The new publication *Advanced Ceramic Materials* is created.

1985 Paul Holbrook is named executive director.

1986 New headquarters at Brooksedge have a grand opening December 4.

1986 The society's building at 65 Ceramic Drive is sold.

1986 A Society delegation goes to France and West Germany.

1987 Society delegation goes to South Korea and England.

1988 Society delegation goes to Australia for Austceram '88 and Spain and Italy.

1988 Proceedings series *Ceramic Transactions* is introduced.

1989 *Advanced Ceramic Materials* is incorporated into the *Journal of the American Ceramic Society.*

1989 The Society holds its First International Ceramic Science and Technology Congress in Anaheim, California, October 31-November 3.

1990 The Society begins an initiative in pre-college education programming.

1990 The Society undertakes a new Strategic Planning initiative.

1991 A Society delegation goes to Japan to celebrate the Ceramic Society of Japan's 100th Anniversary, October 16-17.

1991 The Society holds an open house December 6 at the new headquarters on Ceramic Place in Westerville, Ohio.

1993 The Society holds its first off-shore sponsored meeting, the PAC RIM meeting, November 7-10.

1993 The Ceramic Information Center opens at Society headquarters.

1994 The First Biennial Ceramic Manufacturers & Suppliers Workshop & Exposition is held in Louisville, Kentucky, September 25-28.

1995 The Society launches its World Wide Web site: *www.acers.org*

1995 The Society sends a delegation to China in recognition of the Chinese Ceramic Society's 50th Anniversary Celebration, October 9-13.

1995 The American Ceramic Industry Association (ACIA), the first Society subsidiary, is approved.

1996 The first woman president, Carol Jantzen, is sworn in.

1996 The Society acquires *Ceramics Monthly*.

1997 *Pottery Making Illustrated*, a new publication, publishes its inaugural issue.

1998 The first Ceramics Pavilion is sponsored by the Society at the National Design Engineering Show and Conference in Chicago.

1998 The 100th Annual Meeting takes place in Cincinnati, May 4-7.

W. Paul Holbrook

W. Paul Holbrook has served as ACerS executive director from 1985 to this writing. After a 26-year career in the ceramics industry, he was chosen as executive director from more than 300 applicants. He has been recognized within the Society as a pioneer in implementing sound financial management tools and strategies for the organization. He works closely with the membership to define an objectives-oriented vision for the Society and its future. He is praised for his attention to members' interests and for enhancing the value of membership in the Society.

Until joining the Society, Mr. Holbrook was the works manager for Kaiser Aluminum and Chemical Corp.'s Aluminum Division in Gramercy, Louisiana. In this position, he was responsible for all functions of plant operations, including operating plans, capital spending, overhead budgets and personnel organization. He also worked for Volta Aluminum, a joint company owned by Kaiser and Reynolds Metals Co., which was located in the nation of Ghana. He led Volta through some difficult periods caused by a military coup in Ghana and a government takeover. Other positions he held during his career include work with the Oakland-based company's Refractories Division in both Oakland and Gary, Indiana.

Mr. Holbrook graduated from West Virginia University, where he earned a bachelor's degree in chemical engineering. Mr. Holbrook also attended Harvard University's Business School Program of Management Development in 1965.

PRESIDENTS

1899-1900	H.A. Wheeler
1900-01	Karl Langenbeck
1901-02	C.F. Binns
1902-03	Ernest Mayer
1903-04	E.C. Stover
1904-05	F.W. Walker, Sr.
1905-06	W.D. Gates
1906-07	W.D. Richardson
1907-08	S.G. Burt
1908-09	A.V. Bleininger
1909-10	R.C. Purdy
1910-11	Heinrich Ries
1911-12	Charles Weelans
1912-13	A.S. Watts
1913-14	Ellis Lovejoy
1914-15	C.W. Parmelee
1915-16	R.R. Hice
1916-17	L.E. Barringer
1917-18	G.H. Brown
1918-19	H.F. Staley
1919-20	R.T. Stull
1920-21	R.H. Minton
1921-22	F.K. Pence
1922-23	F.H. Riddle
1923-24	A.F. Greaves-Walker
1924-25	R.D. Landrum
1925-26	E. Ward Tillotson
1926-27	R.L. Clare
1927-28	B. Mifflin Hood
1928-29	M.C. Booze
1929-30	G.A. Bole
1930-31	Edward Orton, Jr.
1931-32	E.V. Eskesen
1932-33	E.P. Poste
1933-34	J.C. Hostetter
1934-35	W. Keith McAfee
1935-36	J.M. McKinley
1936-37	F.C. Flint
1937-38	R.B. Sosman
1938-39	V.V. Kelsey
1939-40	A.I. Andrews
1940-41	J.L. Carruthers
1941-42	J.T. Littleton
1942-43	L.J. Trostel, Sr.
1943-44	C.E. Bales
1944-45	E.H. Fritz
1945-46	C. Forrest Tefft
1946-47	J.E. Hansen
1947-48	J.D. Sullivan
1948-49	J.W. Whittemore
1949-50	H.M. Kraner
1950-51	J.W. Hepplewhite
1951-52	Howard R. Lillie
1952-53	W. Edward Cramer
1953-54	R.R. Danielson
1954-55	Ray W. Pafford
1955-56	Robert Twells
1956-57	Karl Schwartzwalder
1957-58	John F. McMahon
1958-59	R.S. Bradley
1959-60	Oscar G. Burch
1960-61	George H. Spencer-Strong
1961-62	John S. Nordyke
1962-63	John H. Koenig
1963-64	Paul V. Johnson
1964-65	Elburt F. Osborn
1965-66	Howard P. Bonebrake
1966-67	George J. Bair
1967-68	James S. Owens
1968-69	Loran S. O'Bannon
1969-70	Arthur J. Blume
1970-71	J. Earl Frazier
1971-72	William J. Smothers
1972-73	T.J. Planje
1973-74	James R. Johnson
1974-75	Joseph E. Burke
1975-76	Ralston Russell, Jr.
1976-77	Stephen D. Stoddard
1977-78	Lyle A. Holmes
1978-79	John B. Wachtman, Jr.
1979-80	Malcolm G. McLaren
1980-81	William R. Prindle
1981-82	James I. Mueller
1982-83	Robert J. Beals
1983-84	J. Lambert Bates
1984-85	Richard M. Spriggs
1985-86	Edwin Ruh
1986-87	Joseph L. Pentecost
1987-88	Dale E. Niesz
1988-89	William H. Rhodes
1989-90	William H. Payne
1990-91	Robert J. Eagan
1991-92	Dennis Readey
1992-93	George MacZura
1993-94	Richard Tressler
1994-95	David W. Johnson, Jr.
1995-96	Delbert Day
1996-97	Carol M. Jantzen
1997-98	James W. McCauley
1998-99	Stephen W. Freiman

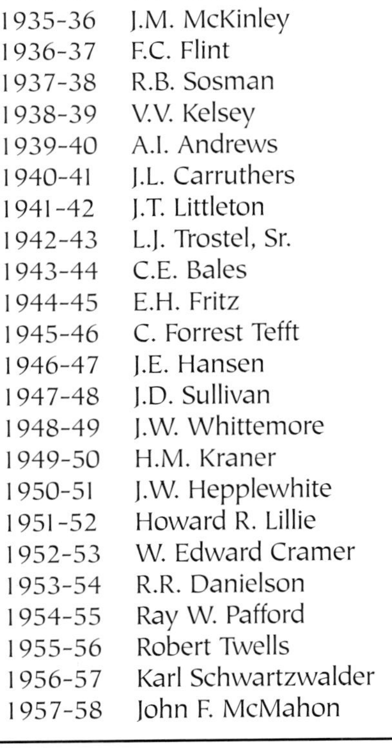

ACKNOWLEDGMENTS

Like so much else in the first century of The American Ceramic Society, this book has depended on a shared effort for its creation. On behalf of the book team, I would like to offer some special recognition for outstanding contributions to this festive history.

The Centennial Celebration Book Committee represents the kind of commitment from members that has clearly made a positive difference for the Society from the beginning. I have seen testament to members' dedication in documents and historical fact, but the Book Committee's work I witnessed myself. Without their continued interest in every facet of the book, it would never have come to be.

They provided insights, research, fact-checking, photographs and proofreading. Our respectful gratitude goes to:

David W. Johnson, Jr., Chair
James W. McCauley, Co-Chair
Margaret Adams-Rasmussen
William H. Payne
William R. Prindle
John B. Wachtman, Jr.
Mark A. Glasper, ACerS Staff Representative

In addition, Paul F. Becher kindly read the final proofs.

All of us who worked on the book are proud of the result, but the true over-arching value of this project has been in the research. As a result of this Centennial recollection, the historical materials at ACerS headquarters have been organized into a working archive. In addition to cataloging papers and photographs, the project has required a great deal of historical analysis and selection. Because the book was planned from the beginning to be a coffee-table book — not a formal history but rather an informal, lighthearted and visually appealing celebration — a great deal of primary research is not recounted in the final volume. However, it exists in the archives now, ready for later researchers and historians.

Linda Lakemacher, a former ACerS publications director, has been responsible for this important work, providing the guiding intelligence and experience, as well as the physical and mental endurance required. Her long familiarity with and affection for everything ACerS has brought a level of accuracy and insight to the archives and the book research that would have been impossible from a stranger. This book and the archives are a tribute to her and, in a lasting way, to her years with the Society that inspired her dedication.

We are all more appreciative than simple acknowledgments can reflect to the members who were so willing to be interviewed for this book, providing perspectives on the Society that have been so valuable:

Alice Alexander
Neil N. Ault
J. Lambert Bates
Robert J. and Lois Beals
Morris Berg
Seymour L. Blum
Denis A. Brosnan
Joseph E. Burke
John A. Coppola
William B. Crandall
Delbert E. Day
Winston Duckworth
Richard A. Eppler
Diane C. Folz
Van Derck Frechette
Donald E. Frith
Geoffrey J. Frohnsdorff
Hans Hausner
J. Raymond Hensler
Arthur H. Heuer
Carol M. Jantzen
James R. Johnson
Charles Kuen Kao
Charles R. Kurkjian
Harry H. Linden
Robert D. Maurer
John B. MacChesney
William S. Mills
Suzanne R. Nagel
Jack S. Nordyke
William H. Payne

L. David Pye
Dennis W. Readey
Merle D. Rigterink
Della M. Roy
Rustum Roy
Edwin and Bette Ruh
Ralston Russell, Jr.
Samuel J. and Joan Schneider
J. Richard Schorr
Richard M. Spriggs
Harold W. Stetson
Stephen D. Stoddard
Thomas G. Stoebe
Donald S. Stookey
Richard E. Tressler
Louis J. Trostel, Jr.
John B. and Edith Wachtman, Jr.
Russell K. Wood

We are thankful to the following for providing photos and other memorabilia:

Hans Hausner
Winston Duckworth
Diane C. Folz
John J. Petrovic
Edwin and Bette Ruh
Samuel J. and Joan Schneider
Vishwa N. Shukla
Louis J. Trostel, Jr.

Special notes of appreciation are due the following who were very helpful to Linda in her work:

Richard and Susan Hommel for their hospitality, and Richard for his frequent help in locating information and for sharing his insights on the Society and the industry, particularly in the Pittsburgh area.

Ralston "Bruzz" Russell, Jr. for sharing his marvelous memory and spending many hours wading through the Society's photo archives, providing "name, rank and serial number" for so many of the previously unidentified faces.

Richard Schorr for organizing and searching the Orton Foundation Archives and for his many insights on Edward Orton, Jr. and his contributions to the Society and to the industry.

The Society headquarters staff have been patient and generous in their cooperation throughout the research and writing phases of the project.

Special thanks are due to Greg Geiger, Thomas Shreves and Terry Fogle of The American Ceramic Society's Ceramic Information Center, for their hospitality and their willing assistance in locating and evaluating archive and library materials. Greg Geiger's orientation to ceramics terminology early in the project (which he was too polite to call "Ceramics for Dummies") helped me get the needed focus to begin.

And, on a personal note, I would like to express my own thanks to:

Paul Holbrook, who shepherds the Society's resources with tough bargaining but a gentle heart;
Dave and Bonnie Johnson, whose caring has crossed distance with palpable force at just the right moments during challenging times. (Bell may have provided the technology, but the spirit is uniquely yours!);
Linda and Bob Lakemacher for the serendipity of their friendship, their lovely home and garden, and for the very happy memories of lively talk and laughter over too much to eat; and Linda especially for rising to the unexpected and carrying this project;
And, of course, Joe Zeller, who introduced me to The American Ceramic Society in the first place. ▲

Jane Mobley

PHOTOGRAPHS AND ILLUSTRATIONS

A picture makes a thousand memories for people who have shared building an organization. Any celebration of a century of ACerS' achievements would be insufficient without visual images to trigger the individual recollections that will enrich this volume for each reader. However, no selections can be complete. Inevitably, each reader will recall faces or events that might well have been pictured, but are not. For that, the book team offers its apologies.

Photographs and illustrations for this book came almost entirely from five sources:

- The American Ceramic Society archives;
- work commissioned by ACerS for this volume;
- corporate members of ACerS;
- universities; and
- individual members.

The materials owned by ACerS include a remarkable collection of approximately 300 glass slides, made or obtained between about 1880 and 1910 by Edward Orton, Jr., and now held in the Ross C. Purdy Museum of Ceramics. These slides depict ceramic materials companies and factories from this period, mostly in the northeastern United States, although other geographic areas are represented. They are available for research and are partially catalogued and some have been made into photographic negatives.

Ellen Dallager is a Columbus, Ohio-based photographer who is best known for her work with professional associations and meetings nationwide, including ACerS. Many members know Ellen from annual and division meetings. This book provided an opportunity to showcase some of Ellen's other talents, such as the photographs of ceramic objects that lend this book much of its style.

Materials held in the ACerS archives are available for research at headquarters in Westerville, Ohio. Anyone who would like to examine these materials should make arrangements with ACerS staff in advance of a visit, as one would expect to do with corporate archives. While the ACerS archives are well organized and documented, they are not part of the James Mueller Library and are not open to the public on a daily basis.

All corporate members of ACerS were invited to participate in this project by notices in Society publications and by letter. At least one photograph or illustration was used in the book from each company that responded. We are especially grateful to these companies; without their help it would have been impossible to illustrate the wonderful reach of the ceramics industry.

Universities with ceramics departments or significant programs in materials sciences departments were invited to send photographs, and the ones that responded are represented in the book, with appreciation to the faculty and staff who helped us located and identify these pictures.

Similarly, all members of ACerS were invited to send photographs or memorabilia, and at least one example was used from every person who sent materials. Quite a number of members have kept scrapbooks; some of these were — understandably — too cumbersome or too precious for their owners to send, but some are represented here. Everyone involved in making this book is especially thankful to the members who shared their photographs and accompanying memories. The snapshots of friends and colleagues together at various events or on excursions probably do the most to capture the wonderful spirit of this organization. ▲

The Ohio State University: *elementary schoolchildren, 138.*

O'Keefe Ceramics: *man at drafting table, 4; ceramic templates, 14; ceramic products, 163.*

Old Hickory Clay Company: *late 1930s air floated ball clay bagging operation, 145.*

The Pennsylvania State University, Department of Materials Science and Engineering: *thermal expansion recorder, 63; corrosion testing, 126; Bob Grurer, 130; Keramos chapter 1952, 132; lab equipment of late 1940s, 132; loading ceramics, 133; students using SIM machine, 137; 1953 Ceramic open house, 138.*

The Pfaltzgraff Co.: *historical product, 143; old pottery, 144.*

PhotoDisc: *fiber optics, xii; sand, 87, 127.*

Copyright/Pittsburgh Post-Gazette, 1998 all rights reserved. Reprinted with permission: *Monongahela House, 20, 169; Bedford Springs Hotel, 90.*

PPG Industries, Inc.: *stained-glass window, 1; rolling plate glass, 34; polishing plate glass, 152.*

Crystar® vertical wafer boats, 141; Crystar® wafer boats, 155.

Schneider, Joan and Samuel J.: *ACerS 71st Annual Meeting and Exposition program, 49; Materials and Equipment Division and Whitewares 1976, 99; members dancing, 99; Bedford Springs luncheon, 99.*

Schott Glass Technologies, Inc.: *Zeroder, 8; Ceran glass-ceramic cooktops, 153; glass rods, 155.*

Sociedad Mexicana De Ceramica A.C.: *Lamosa Tile Plant in Monterrey, 109; announcement of the inaugural Mal McLaren address, 110.*

Toyota Central R&D Labs, Inc.: *spin test, 161.*

Tressler, Richard E.: *phase diagram, 3.*

Trostel, L.J., Jr.: *gala dinner, 86; Bales as Toastmaster, 89; photo of Trostel, 90.*

United States Nuclear Regulatory Commission: *D.C. Cook Plant, Michigan, 103.*

University of Illinois at Urbana-Champaign, Department of Materials Science and Engineering: *March 1997, students at annual engineering open house, 139.*

...ty of Washington, Materials Science and ...ing: *Bonnie J. Dunbar, 43; dental ceramics dis- ...; James I. Mueller, 134; X-ray lab 1950, 135; ...nd Robert J. Campbell, 138.*

..., Pamela, Smithsonian Institution: *fertility ...*

... McDanel: *man with tapalog, 10; woman at ... 13; men in brick factory, 15; women hand-rolling ...edia, 1955, 47.*

The American Ceramic Society
100 Years

was digitally composed in ITC Tiepolo and Americana
and printed on Fortune Gloss Book.

Created by Highwater Editions.